Arthur Latham Baker

Elliptic functions

An elementary text-book for students of mathematics

Arthur Latham Baker

Elliptic functions

An elementary text-book for students of mathematics

ISBN/EAN: 9783337277055

Printed in Europe, USA, Canada, Australia, Japan

Cover: Foto ©berggeist007 / pixelio.de

More available books at **www.hansebooks.com**

ELLIPTIC FUNCTIONS.

AN ELEMENTARY TEXT-BOOK FOR STUDENTS OF MATHEMATICS.

BY

ARTHUR L. BAKER, C.E., Ph.D.,

Professor of Mathematics in the Stevens School of the Stevens Institute of Technology, Hoboken. N. J.; formerly Professor in the Pardee Scientific Department, Lafayette College, Easton, Pa.

$$\sin am\ u = \frac{1}{\sqrt{k}} \cdot \frac{H(u)}{\Theta(u)}.$$

NEW YORK:
JOHN WILEY & SONS,
53 East Tenth Street.
1890.

PREFACE.

In the works of Abel, Euler, Jacobi, Legendre, and others, the student of Mathematics has a most abundant supply of material for the study of the subject of Elliptic Functions.

These works, however, are not accessible to the general student, and, in addition to being very technical in their treatment of the subject, are moreover in a foreign language.

It is in the hope of smoothing the road to this interesting and increasingly important branch of Mathematics, and of putting within reach of the English student a tolerably complete outline of the subject, clothed in simple mathematical language and methods, that the present work has been compiled.

New or original methods of treatment are not to be looked for. The most that can be expected will be the simplifying of methods and the reduction of them to such as will be intelligible to the average student of Higher Mathematics.

I have endeavored throughout to use only such methods as are familiar to the ordinary student of Calculus, avoiding those methods of discussion dependent upon the properties of double periodicity, and also those depending upon Functions of Complex Variables. For the same reason I have not carried the discussion of the Θ and H functions further.

Among the minor helps to simplicity is the use of zero subscripts to indicate decreasing series in the Landen Transformation, and of numerical subscripts to indicate increasing series. I have adopted the notation of Gudermann, as being more simple than that of Jacobi.

I have made free use of the following works: JACOBI'S Fundamenta Nova Theoriæ Func. Ellip.; HOUEL'S Calcul Infinitésimal; LEGENDRE'S Traité des Fonctions Elliptiques; DUREGE'S Theorie der Elliptischen Functionen; HERMITE'S Théorie des Fonctions Elliptiques; VERHULST'S Théorie des Functions Elliptiques; BERTRAND'S Calcul Intégral; LAURENT'S Théorie des Fonctions Elliptiques; CAYLEY'S Elliptic Functions; BYERLY'S Integral Calculus; SCHLOMILCH'S Die Höheren Analysis; BRIOT ET BOUQUET'S Fonctions Elliptiques.

I have refrained from any reference to the Gudermann or Weierstrass functions as not within the scope of this work, though the Gudermannians might have been interesting examples of verification formulæ. The arithmetico-geometrical mean, the march of the functions, and other interesting investigations have been left out for want of room.

CONTENTS.

	PAGE
INTRODUCTORY CHAPTER,	1
CHAP. I. ELLIPTIC INTEGRALS,	4
II. ELLIPTIC FUNCTIONS,	16
III. PERIODICITY OF THE FUNCTIONS,	22
IV. LANDEN'S TRANSFORMATION,	30
V. COMPLETE FUNCTIONS,	45
VI. EVALUATION FOR ϕ,	48
VII. FACTORIZATION OF ELLIPTIC FUNCTIONS,	51
VIII. THE Θ FUNCTION,	66
IX. THE Θ AND H FUNCTIONS,	69
X. ELLIPTIC INTEGRALS OF THE SECOND ORDER,	81
XI. ELLIPTIC INTEGRALS OF THE THIRD ORDER,	90
XII. NUMERICAL CALCULATIONS, q,	94
XIII. NUMERICAL CALCULATIONS, K',	98
XIV. NUMERICAL CALCULATIONS, u,	102
XV. NUMERICAL CALCULATIONS, ϕ,	108
XVI. NUMERICAL CALCULATIONS, $E(k, \phi)$,	111
XVII. APPLICATIONS,	115

ELLIPTIC FUNCTIONS.

INTRODUCTORY CHAPTER.*

THE first step taken in the theory of Elliptic Functions was the determination of a relation between the amplitudes of three functions of either order, such that there should exist an algebraic relation between the three functions themselves of which these were the amplitudes. It is one of the most remarkable discoveries which science owes to Euler. In 1761 he gave to the world the complete integration of an equation of two terms, each an elliptic function of the first or second order, not separately integrable.

This integration introduced an arbitrary constant in the form of a third function, related to the first two by a given equation between the amplitudes of the three.

In 1775 Landen, an English mathematician, published his celebrated theorem showing that any arc of a hyperbola may be measured by two arcs of an ellipse, an important element of the theory of Elliptic Functions, but *then* an isolated result. The great problem of comparison of Elliptic Functions of different moduli remained unsolved, though Euler, in a measure, exhausted the comparison of functions of the same modulus. It was completed in 1784 by Lagrange, and for the computation

---

* Condensed from an article by Rev. Henry Moseley, M.A., F.R.S., Prof. of Nat. Phil. and Ast., King's College, London.

of numerical results leaves little to be desired. The value of a function may be determined by it, in terms of increasing or diminishing moduli, until at length it depends upon a function having a modulus of zero, or unity.

For all practical purposes this was sufficient. The enormous task of calculating tables was undertaken by Legendre. His labors did not end here, however. There is none of the discoveries of his predecessors which has not received some perfection at his hands; and it was he who first supplied to the whole that connection and arrangement which have made it an independent science.

The theory of Elliptic Integrals remained at a standstill from 1786, the year when Legendre took it up, until the year 1827, when the second volume of his Traité des Fonctions Elliptiques appeared. Scarcely so, however, when there appeared the researches of Jacobi, a Professor of Mathematics in Königsberg, in the 123d number of the Journal of Schumacher, and those of Abel, Professor of Mathematics at Christiania, in the 3d number of Crelle's Journal for 1827.

These publications put the theory of Elliptic Functions upon an entirely new basis. The researches of Jacobi have for their principal object the development of that general relation of functions of the first order having different moduli, of which the scales of Legrange and Legendre are particular cases.

It was to Abel that the idea first occurred of treating the Elliptic Integral as a function of its amplitude. Proceeding from this new point of view, he embraced in his speculations all the principal results of Jacobi. Having undertaken to develop the principle upon which rests the fundamental proposition of Euler establishing an algebraic relation between three functions which have the same moduli, dependent upon a certain relation of their amplitudes, he has extended it from three to an indefinite number of functions; and from Elliptic Functions to an infinite number of other functions embraced under an indefinite number of classes, of which that of Elliptic Functions is but one ; and each class having a division analogous to that of Elliptic Functions into three orders having common properties.

The discovery of Abel is of infinite moment as presenting the first step of approach towards a more complete theory of the infinite class of ultra elliptic functions, destined probably ere long to constitute one of the most important of the branches of transcendental analysis, and to include among the integrals of which it effects the solution some of those which at present arrest the researches of the philosopher in the very elements of physics.

CHAPTER I.

ELLIPTIC INTEGRALS.

THE integration of irrational expressions of the form

$$Xdx\sqrt{A + Bx + Cx^2},$$

or

$$\frac{Xdx}{\sqrt{A + Bx + Cx^2}},$$

X being a rational function of x, is fully illustrated in most elementary works on Integral Calculus, and shown to depend upon the transcendentals known as logarithms and circular functions, which can be calculated by the proper logarithmic and trigonometric tables.

When, however, we undertake to integrate irrational expressions containing higher powers of x than the square, we meet with insurmountable difficulties. This arises from the fact that the integral sought depends upon a new set of transcendentals, to which has been given the name of *elliptic functions*, and whose characteristics we will learn hereafter.

The name of Elliptic Integrals has been given to the simple integral forms to which can be reduced all integrals of the form

(1) $$V = \int F(X, R)dx,$$

where $F(X, R)$ designates a rational function of x and R, and R represents a radical of the form

$$R = \sqrt{Ax^4 + Bx^3 + Cx^2 + Dx + E},$$

ELLIPTIC INTEGRALS.

where A, B, C, D, E indicate constant coefficients.

We will show presently that all cases of Eq. (1) can be reduced to the three typical forms

$$\text{(2)} \quad \begin{aligned} & \int_0^x \frac{dx}{\sqrt{(1-x^2)(1-k^2x^2)}}, \\ & \int_0^x \frac{x^2 dx}{\sqrt{(1-x^2)(1-k^2x^2)}}, \\ & \int_0^x \frac{dx}{(x^2+a)\sqrt{(1-x^2)(1-k^2x^2)}}, \end{aligned}$$

which are called elliptic integrals of the first, second, and third order.

Why they are called *Elliptic* Integrals we will learn further on. The transcendental functions which depend upon these integrals, and which will be discussed in Chapter IV, are called *Elliptic Functions*.

The most general form of Eq. (1) is

$$\text{(3)} \quad V = \int \frac{A + BR}{C + DR} dx;$$

where $A, B, C,$ and D stand for rational integral functions of x.

$\dfrac{A + BR}{C + DR}$ can be written

$$\frac{A+BR}{C+DR} = \frac{AC - BDR^2}{C^2 - D^2R^2} - \frac{(AD-CB)R^2}{C^2 - D^2R^2} \cdot \frac{1}{R}$$

$$= N - \frac{P}{R};$$

N and P being rational integral functions of x. Whence Eq. (3) becomes

(4) $$V = \int N dx - \int \frac{P dx}{R}.$$

Eq. (4) shows that the most general form of V can be made to depend upon the expressions

(5) $$V' = \int \frac{P dx}{R},$$

and
$$\int N dx.$$

This last form is rational, and needs no discussion here. We can write

$$P = \frac{G_0 + G_1 x + G_2 x^2 + \cdots}{H_0 + H_1 x + H_2 x^2 + \cdots}$$

$$= \frac{G_0 + G_2 x^2 + G_4 x^4 + \cdots + (G_1 + G_3 x^2 + \cdots) x}{H_0 + H_2 x^2 + H_4 x^4 + \cdots + (H_1 + H_3 x^2 + \cdots) x}.$$

Multiplying both numerator and denominator by

$$H_0 + H_2 x^2 + H_4 x^4 + \cdots - (H_1 + H_3 x^2 + H_5 x^4 + \cdots) x,$$

we have a new numerator which contains only powers of x^2. The result takes the following form:

$$P = \frac{M_0 + M_2 x^2 + M_4 x^4 + \cdots + (M_1 + M_3 x^2 + M_5 x^4 + \cdots) x}{N_0 + N_2 x^2 + N_4 x^4 + N_6 x^6 + \cdots}$$

$$= \Phi(x^2) + \Psi(x^2) \cdot x.$$

Equation (5) thus becomes

(6) $$V' = \int \frac{\Phi(x^2) dx}{R} + \int \frac{\Psi(x^2) \cdot x \cdot dx}{R}.$$

We shall see presently that R can always be assumed to be of the form

$$\sqrt{(1-x^2)(1-k^2x^2)}.$$

Therefore, putting $x^2 = z$, the second integral in Eq. (6) takes the form

$$\frac{1}{2}\int \frac{\Psi(z) \cdot dz}{\sqrt{(1-z)(1-k^2z)}},$$

which can be integrated by the well-known methods of Integral Calculus, resulting in logarithmic and circular transcendentals.

There remains, therefore, only the form

$$\int \frac{\Phi(x^2)dx}{R}$$

to be determined.

We will now show that R can always be assumed to be in the form

$$\sqrt{(1-x^2)(1-k^2x^2)}.$$

We have

$$R = \sqrt{Ax^4 + Bx^3 + Cx^2 + Dx + E}$$
$$= \sqrt{G(x-a)(x-b)(x-c)(x-d)},$$

a, b, c, and d being the roots of the polynomial of the fourth degree, and G any number, real or imaginary, depending upon the coefficients in the given polynomial.

Substituting in equation (1)

$$x = \frac{p+qy}{1+y},$$

we have

(7) $$V = \int \phi(y, \rho)dy,$$

8   ELLIPTIC FUNCTIONS.

ρ designating the radical

$$\rho = \sqrt{G[p-a+(q-a)y][p-b+(q-b)y][p-c+(q-c)y]}\ldots$$

In order that the odd powers of y under the radical may disappear we must have their coefficients equal to zero; i.e.,

$$(p-a)(q-b)+(p-b)(q-a) = 0,$$
$$(p-c)(q-d)+(p-d)(q-c) = 0;$$

whence

$$2pq - (p+q)(a+b) + 2ab = 0,$$
$$2pq - (p+q)(c+d) + 2cd = 0,$$

and

(8)
$$\begin{cases} pq = \dfrac{ab(c+d) - cd(a+b)}{a+b-(c+d)}, \\ p+q = \dfrac{2ab - 2cd}{a+b-(c+d)}. \end{cases}$$

Equation (8) shows that p and q are real quantities, whether the roots a, b, c, and d are real or imaginary; a, b, and c, d being the conjugate pairs.

Hence equation (1) can always be reduced to the form of equation (7), which contains only the second and fourth powers of the variable.

This transformation seems to fail when $a+b-(c+d) = 0$; but in that case we have

$$R = \sqrt{G[x^2 - (a+b)x + ab][x^2 - (a+b)x + cd]},$$

and substituting

$$x = y - \frac{a+b}{2}$$

will cause the odd powers of y to disappear as before.

If the radical should have the form

$$\sqrt{G(x-a)(x-b)(x-c)},$$

placing $x = y^2 + a$, we get

$$V = \int \phi(y, \rho)dy,$$

$$\rho = \sqrt{G(y^2 + a - b)(y^2 + a - c)},$$

ϕ designating a rational function of y and ρ.

Thus all integrals of the form contained in equation (1), in which R stands for a quadratic surd of the third or fourth degree, can be reduced to the form

(9) $$V = \int \phi(x, R)dx,$$

R designating a radical of the form

$$\sqrt{G(1 + mx^2)(1 + nx^2)},$$

m and n designating constants.

It is evident that if we put

$$x' = x\sqrt{-m}, \quad k^2 = -\frac{n}{m},$$

we can reduce the radical to the form

$$\sqrt{(1 - x^2)(1 - k^2x^2)}.$$

We shall see later on that the quantity k^2, to which has been given the name *modulus*, can always be considered real and less than unity.

Combining these results with equation (6), we see that the integration of equation (1) depends finally upon the integration of the expression

(10) $$V'' = \int \frac{\phi(x^2)dx}{\sqrt{(1 - x^2)(1 - k^2x^2)}} = \int \frac{\phi(x^2)dx}{R}.$$

The most general form of $\phi(x^2)$ is

$$\phi(x^2) = \frac{M_0 + M_2 x^2 + M_4 x^4 + \cdots}{N_0 + N_2 x^2 + N_4 x^4 + \cdots}$$

$$= P_0 + P_2 x^2 + P_4 x^4 + P_6 x^6 + \cdots$$

$$+ \Sigma \frac{L}{(x^2 + a)^n}.$$

Hence

(11) $$V'' = \Sigma P \int \frac{x^{2m} dx}{R} + \Sigma L \int \frac{dx}{(x^2 + a)^n R}.$$

But $\int \frac{x^{2m} dx}{R}$ depends upon $\int \frac{dx}{R}$ and $\int \frac{x^2 dx}{R}$, which can be shown as follows:

Differentiating Rx^{2m-3}, we have

$$d[x^{2m-3} R] = d\left[x^{2m-3} \sqrt{\alpha + \beta x^2 + \gamma x^4}\right]$$
$$= (2m - 3) x^{2m-4} dx \sqrt{\alpha + \beta x^2 + \gamma x^4}$$
$$+ \frac{x^{2m-3}(\beta x + 2\gamma x^3) dx}{\sqrt{\alpha + \beta x^2 + \gamma x^4}}.$$

Integrating and collecting, we get

$$R x^{2m-3} = (2m-3)\alpha \int \frac{x^{2m-4} dx}{R} + (2m-2)\beta \int \frac{x^{2m-2} dx}{R}$$
$$+ (2m-1)\gamma \int \frac{x^{2m} dx}{R}$$

(12) $$= \alpha' \int \frac{x^{2m-4} dx}{R} + \beta' \int \frac{x^{2m-2} dx}{R} + \gamma' \int \frac{x^{2m} dx}{R}.$$

Whence we get, by taking $m = 2$,

(13) $$Rx = \alpha \int \frac{dx}{R} + \beta \int \frac{x^2 dx}{R} + \gamma \int \frac{x^4 dx}{R},$$

which shows that the general expression $\int \frac{x^{2m} dx}{R}$ can be found by successive calculations, when we are able to integrate the expressions

$$\int \frac{dx}{R} \quad \text{and} \quad \int \frac{x^2 dx}{R},$$

the first and second of equation (2).

We will now consider the second class of terms in eq. (11), viz., $\frac{Ldx}{(x^2 + a)^n R}$.

This second term is as follows:

(14) $$\Sigma \int \frac{L}{(x^2 + a)^n R} = \int \frac{A dx}{(x^2 + a)^n R} + \int \frac{B dx}{(x^2 + a)^{n-1} R}$$
$$+ \int \frac{C dx}{(x^2 + a)^{n-2} R} + \cdots$$

Each of these terms can be shown to depend ultimately upon terms of the form

$$\frac{x^2 dx}{R}, \quad \frac{dx}{R}, \quad \text{and} \quad \frac{dx}{(x^2 + a)R}.$$

The two former will be recognized as the two ultimate forms already discussed, the first and second of equation (2). The third is the third one of equation (2).

This dependence of equation (14) can be shown as follows:

We have

$$d\left[\frac{xR}{(x^2+a)^{n-1}}\right] = \frac{(x^2+a)^{n-1}(xdR+Rdx)-2x^2R(n+1)(x^2+a)^{n-2}dx}{(x^2+a)^{2n-2}}$$
$$= \frac{(x^2+a)(xdR+Rdx)-2x^2R(n-1)dx}{(x^2+a)^n}.$$

Substituting the value of

$$R = \sqrt{\alpha+\beta x^2+\gamma x^4} \quad \text{and} \quad dR = (\beta x + 2\gamma x^3)\frac{dx}{R},$$

we get

$$d\left[\frac{xR}{(x^2+a)^{n-1}}\right] =$$

$$\frac{(x^2+a)(\beta x^2+2\gamma x^4+\alpha+\beta x^2+\gamma x^4)-2x^2(n-1)(\alpha+\beta x^2+\gamma x^4)}{(x^2+a)^n} \cdot \frac{dx}{R}$$

$$= \frac{\begin{array}{l|l|l|l}3\gamma & x^6 & +2\beta & x^4 & +2a\beta & x^2 & +a\alpha \\ -2(n-1)\gamma & & +3a\gamma & & +\alpha & & \\ & & -2(n-1)\beta & & -2(n-1)\alpha & & \end{array}}{(x^2+a)^n} \cdot \frac{dx}{R}$$

$$= \frac{\begin{array}{l|l|l|l}-(2n-5)\gamma x^6 & -(2n-4)\beta & x^4 & -(2n-3)\alpha & x^2 & +a\alpha \\ & +3a\gamma & & -2a\beta & & \end{array}}{(x^2+a)^n} \cdot \frac{dx}{R};$$

or, by substituting in the numerator $x^2 = z - a$,

$$= \frac{\begin{array}{l|l|l}-(2n-5)\gamma z^3 + (2n-5)3a\gamma & z^2 - (2n-5)3a^2\gamma & z + (2n-5)a^3\gamma \\ -(2n-4)\beta & +(2n-4)2a\beta & -(2n-4)a^2\beta \\ +3a\gamma & -6a^2\gamma & +3a^2\gamma \\ & -(2n-3)\alpha & +(2n-3)a\alpha \\ & +2a\beta & -2a^2\beta \\ & & +a\alpha \end{array}}{(x^2+a)^n} \cdot \frac{dx}{R};$$

or, after resubstituting $z = x^2 + a$, and integrating,

(15) $\dfrac{xR}{(x^2+a)^{n-1}} = -(2n-5)\gamma \displaystyle\int \dfrac{dx}{(x^2+a)^{n-3}R}$

$\phantom{(15)\dfrac{xR}{(x^2+a)^{n-1}}=} -(2n-4)(\beta - 3a\gamma) \displaystyle\int \dfrac{dx}{(x^2+a)^{n-2}R}$

$\phantom{(15)\dfrac{xR}{(x^2+a)^{n-1}}=} -(2n-3)(3a^2\gamma - 2a\beta + \alpha) \displaystyle\int \dfrac{dx}{(x^2+a)^{n-1}R}$

$\phantom{(15)\dfrac{xR}{(x^2+a)^{n-1}}=} +(2n-2)(a^3\gamma - a^2\beta + a\alpha) \displaystyle\int \dfrac{dx}{(x^2+a)^{n}R}.$

$= \alpha_1 \displaystyle\int \dfrac{dx}{(x^2+a)^{n-3}R} + \beta_1 \displaystyle\int \dfrac{dx}{(x^2+a)^{n-2}R} + \gamma_1 \displaystyle\int \dfrac{dx}{(x^2+a)^{n-1}R}$

$\phantom{= \alpha_1 \int \dfrac{dx}{(x^2+a)^{n-3}R}} + \delta_1 \displaystyle\int \dfrac{dx}{(x^2+a)^{n}R}.$

Making $n = 2$, we have

(16) $\dfrac{xR}{(x^2+a)^{-1}} = \alpha_1 \displaystyle\int \dfrac{(x^2+a)dx}{R} + \beta_1 \displaystyle\int \dfrac{dx}{R} + \gamma_1 \displaystyle\int \dfrac{dx}{(x^2+a)R}$

$\phantom{(16) \dfrac{xR}{(x^2+a)^{-1}} =} + \delta_1 \displaystyle\int \dfrac{dx}{(x^2+a)^2 R}.$

Equation (16) shows that

$$\int \dfrac{dx}{(x^2+a)^2 R}$$

depends upon the three forms

$$\int \dfrac{x^2 dx}{R}, \quad \int \dfrac{dx}{R}, \quad \text{and} \quad \int \dfrac{dx}{(x^2+a)R}.$$

the three types of equation (2), and equation (15) shows that the general form

$$\int \frac{dx}{(x^2+a)^n R}$$

depends ultimately upon the same three types.

We have now discussed every form which the general equation (1) can assume, and shown that they all depend ultimately upon one or more of the three types contained in equation (2).

These three types are called the three Elliptic Integrals of the first, second, and third kind, respectively.

Legendre puts $x = \sin \phi$, and reduces the three integrals to the following forms:

(17) $\quad F(k, \phi) = \int_0^\phi \frac{d\phi}{\sqrt{1 - k^2 \sin^2 \phi}};$

$$\frac{1}{k^2} \int_0^\phi \frac{d\phi}{\sqrt{1 - k^2 \sin^2 \phi}} - \frac{1}{k^2} \int_0^\phi \sqrt{1 - k^2 \sin^2 \phi} \cdot d\phi;$$

(18) $\quad \Pi(n, k, \phi) = \int_0^\phi \frac{d\phi}{(1 - n \sin^2 \phi)\sqrt{1 - k^2 \sin^2 \phi}};$

the first being Legendre's integral of the first kind; the form

(19) $\quad E(k, \phi) = \int_0^\phi \sqrt{1 - k^2 \sin^2 \phi} \cdot d\phi$

being the integral of the second kind; and the third one being the integral of the third kind.

The form of the integral of the second kind shows why they are called Elliptic Integrals, the arc of an elliptic quadrant being equal to

$$a \int_0^{\frac{\pi}{2}} \sqrt{1 - e^2 \sin^2 \phi} \cdot d\phi,$$

ϕ being the complement of the eccentric angle.

By easy substitutions, we get from Eqs. (17), (18), and (19) the following solutions:

$$\int_0^\phi \frac{\sin^2 \phi}{\Delta} d\phi = \frac{F - E}{k^2};$$

$$\int_0^\phi \frac{\cos^2 \phi}{\Delta} d\phi = \frac{E - (1 - k^2)F}{k^2};$$

$$\int_0^\phi \frac{\tan^2 \phi}{\Delta} d\phi = \frac{\Delta \tan \phi - E}{1 - k^2};$$

$$\int_0^\phi \frac{\sec^2 \phi}{\Delta} d\phi = \frac{\Delta \tan \phi + (1 - k^2)F - E}{1 - k^2};$$

$$\int_0^\phi \frac{1}{\Delta^3} d\phi = \frac{1}{1 - k^2}\left(E - \frac{k^2 \sin \phi \cos \phi}{\Delta}\right);$$

$$\int_0^\phi \frac{\sin^2 \phi}{\Delta^3} d\phi = \frac{1}{1 - k^2}\left(\frac{E - (1 - k^2)F}{k^2} - \frac{\sin \phi \cos \phi}{\Delta}\right);$$

$$\int_0^\phi \frac{\cos^2 \phi}{\Delta^3} d\phi = \frac{F - E}{k^2} + \frac{\sin \phi \cos \phi}{\Delta}.$$

CHAPTER II.

ELLIPTIC FUNCTIONS.

LET
$$u = \int_0^{\phi} \frac{d\phi}{\sqrt{1 - k^2 \sin^2 \phi}}.$$

ϕ* is called the *amplitude* corresponding to the *argument u*, and is written

$$\phi = \text{am}(u, k) = \text{am } u.$$

The quantity k is called the *modulus*, and the expression $\sqrt{1 - k^2 \sin^2 \phi}$ is written *

$$\sqrt{1 - k^2 \sin^2 \phi} = \Delta \text{ am } u = \Delta\phi,$$

and is called the *delta function* of the amplitude o. u, or *delta of ϕ*, or simply *delta ϕ*.

u can be written

$$u = F(k, \phi).$$

The following abbreviations are used:

$$\sin \phi = \sin \text{ am } u = \text{sn}\dagger u;$$
$$\cos \phi = \cos \text{ am } u = \text{cn}\dagger u;$$
$$\Delta\phi = \Delta \text{ am } u = \text{dn}\dagger u = \Delta u;$$
$$\tan \phi = \tan \text{ am } u = \text{tn } u.$$

Let ϕ and ψ be any two arbitrary angles, and put

$$\phi = \text{am } u;$$
$$\psi = \text{am } v.$$

---

* *Legendre.*
† *Gudermann*, in his "Theorie der Modularfunctionen": Crelle's Journal, Bd. 18.

ELLIPTIC FUNCTIONS.

In the spherical triangle ABC we have from Trigonometry, c and C being constant,

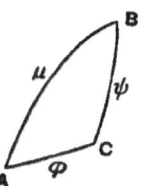

$$\frac{d\phi}{\cos B} + \frac{d\psi}{\cos A} = 0.$$

Since C and c are constant, denoting by k an arbitrary constant, we have

(1) $$\frac{\sin C}{\sin \mu} = k.$$

But

$$\sin A = \sin \psi \frac{\sin B}{\sin \phi} = \sin \psi \frac{\sin C}{\sin \mu} = k \sin \psi.$$

Whence

$$\cos A = \sqrt{1 - \sin^2 A} = \sqrt{1 - k^2 \sin^2 \psi}.$$

In the same manner

$$\cos B = \sqrt{1 - \sin^2 B} = \sqrt{1 - k^2 \sin^2 \phi}.$$

Substituting these values, we get

(2) $$\frac{d\phi}{\sqrt{1 - k^2 \sin^2 \phi}} + \frac{d\psi}{\sqrt{1 - k^2 \sin^2 \psi}} = 0.$$

Integrating this, there results

(3) $$\int_0^\phi \frac{d\phi}{\sqrt{1 - k_2 \sin^2 \phi}} + \int_0^\psi \frac{d\psi}{\sqrt{1 - k^2 \sin^2 \psi}} = \text{const.}$$

When $\phi = 0$, we have $\psi = \mu$, and therefore the constant must be of the form

$$\int_0^\mu \frac{d\phi}{\sqrt{1 - k^2 \sin^2 \phi}},$$

whence

$$(4) \int_0^\phi \frac{d\phi}{\sqrt{1-k^2\sin^2\phi}} + \int_0^\psi \frac{d\psi}{\sqrt{1-k^2\sin^2\psi}} = \int_0^\mu \frac{d\phi}{\sqrt{1-k^2\sin^2\phi}},$$

or

$$u + v = m;$$

and evidently the amplitudes ϕ, ψ, and μ can be considered as the three sides of a spherical triangle, and the relations between the sides of this spherical triangle will be the same as those between ϕ, ψ, and μ.

But the sides of this triangle have imposed upon them the condition

$$\frac{\sin C}{\sin \mu} = k;$$

and since $k < 1$, we must have $\mu > C$, which requires that one of the angles of the triangle shall be obtuse and the other two acute.

In the figure, let C be an acute angle of the triangle ABC, and PQ the equatorial great circle of which C is the pole.

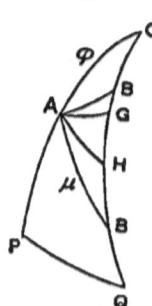

The arc PQ will be the measure of the angle C.

Let AG and AH be the arcs of two great circles perpendicular respectively to CQ and CP. They will of course be shorter than PQ. Hence $AB = \mu$ must intersect CQ in points between CG and HQ, since $\mu > (C = PQ)$. In any case either A or B will be obtuse according as B falls between QH or CG respectively; and the other angle will be acute.

In the case where C is an obtuse angle, it will be easily seen that the angle at A must be acute, since the great circle AD, perpendicular to CP, intersects PQ in D, PD being a quadrant. The same remarks apply to the angle B. Hence, in either case, one of the angles of the triangle is obtuse and the other two are acute, as a result of the condition

$$\frac{\sin C}{\sin \mu} = k < 1.$$

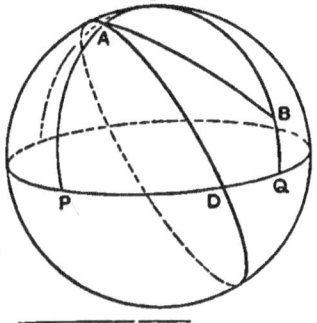

From Trigonometry we have
$\cos \mu = \cos \phi \cos \psi + \sin \phi \sin \psi \cos C;$
and since the angle C is obtuse,

$$\cos C = -\sqrt{1 - \sin^2 C} = -\sqrt{1 - k^2 \sin^2 \mu},$$

and

(5) $\quad \cos \mu = \cos \phi \cos \psi - \sin \phi \sin \psi \sqrt{1 - k^2 \sin^2 \mu},$
the relation sought.

The spherical triangle likewise gives the following relations between the sides:

$$(5)^* \begin{cases} \cos \phi = \cos \mu \cos \psi + \sin \mu \sin \psi \sqrt{1 - k^2 \sin^2 \phi}; \\ \cos \psi = \cos \mu \cos \phi + \sin \mu \sin \phi \sqrt{1 - k^2 \sin^2 \psi}. \end{cases}$$

These give, by eliminating $\cos \mu$,

$$\sin \mu = \frac{\cos^2 \psi - \cos^2 \phi}{\sin \phi \cos \psi \, \varDelta \psi - \sin \psi \cos \phi \, \varDelta \phi};$$

which, after multiplying by the sum of the terms in the denominator and substituting $\cos^2 = 1 - \sin^2$, can be written

$$\sin \mu = \frac{(\sin^2 \phi - \sin^2 \psi)(\sin \phi \cos \psi \, \varDelta \psi + \sin \psi \cos \phi \, \varDelta \phi)}{\sin^2 \phi \cos^2 \psi \, \varDelta^2 \psi - \sin^2 \psi \cos^2 \phi \, \varDelta^2 \phi}.$$

Since the denominator can be written

$$(\sin^2 \phi - \sin^2 \psi)(1 - k^2 \sin^2 \phi \sin^2 \psi),$$

(6) $\quad \sin \mu = \dfrac{\sin \phi \cos \psi \, \varDelta \psi + \sin \psi \cos \phi \, \varDelta \phi}{1 - k^2 \sin^2 \phi \sin^2 \psi}.$

In a similar manner we get

$$(6)^* \begin{cases} \cos \mu = \dfrac{\cos \phi \cos \psi - \sin \phi \sin \psi \, \varDelta \phi \, \varDelta \psi}{1 - k^2 \sin^2 \phi \sin^2 \psi}; \\ \varDelta \mu = \dfrac{\varDelta \phi \, \varDelta \psi - k^2 \sin \phi \sin \psi \cos \phi \cos \psi}{1 - k^2 \sin^2 \phi \sin^2 \psi}. \end{cases}$$

These equations can also be written as follows:

$$(7)\begin{cases}\sin\operatorname{am}(u\pm v)=\dfrac{\sin\operatorname{am}u\cos\operatorname{am}v\,\Delta\operatorname{am}v\pm\sin\operatorname{am}v\cos\operatorname{am}u\,\Delta\operatorname{am}u}{1-k^2\sin^2\operatorname{am}u\sin^2\operatorname{am}v};\\[4pt]\cos\operatorname{am}(u\pm v)=\dfrac{\cos\operatorname{am}u\cos\operatorname{am}v\mp\sin\operatorname{am}u\sin\operatorname{am}v\,\Delta\operatorname{am}u\,\Delta\operatorname{am}v}{1-k^2\sin^2\operatorname{am}u\sin^2\operatorname{am}v};\\[4pt]\Delta\operatorname{am}(u\pm v)=\dfrac{\Delta\operatorname{am}u\,\Delta\operatorname{am}v\mp k^2\sin\operatorname{am}u\sin\operatorname{am}v\cos\operatorname{am}u\cos\operatorname{am}v}{1-k^2\sin^2\operatorname{am}u\sin^2\operatorname{am}v};\end{cases}$$

or

$$(8)\begin{cases}\operatorname{sn}(u\pm v)=\dfrac{\operatorname{sn}u\operatorname{cn}v\operatorname{dn}v\pm\operatorname{sn}v\operatorname{cn}u\operatorname{dn}u}{1-k^2\operatorname{sn}^2u\operatorname{sn}^2v};\\[4pt]\operatorname{cn}(u\pm v)=\dfrac{\operatorname{cn}u\operatorname{cn}v\mp\operatorname{sn}u\operatorname{sn}v\operatorname{dn}u\operatorname{dn}v}{1-k^2\operatorname{sn}^2u\operatorname{sn}^2v};\\[4pt]\operatorname{dn}(u\pm v)=\dfrac{\operatorname{dn}u\operatorname{dn}v\mp k^2\operatorname{sn}u\operatorname{sn}v\operatorname{cn}u\operatorname{cn}v}{1-k^2\operatorname{sn}^2u\operatorname{sn}^2v}.\end{cases}$$

Making $u = v$, we get from the upper sign

$$(9)\begin{cases}\operatorname{sn}2u=\dfrac{2\operatorname{sn}u\operatorname{cn}u\operatorname{dn}u}{1-k^2\operatorname{sn}^4u};\\[4pt]\operatorname{cn}2u=\dfrac{\operatorname{cn}^2u-\operatorname{sn}^2u\operatorname{dn}^2u}{1-k^2\operatorname{sn}^4u}=\dfrac{1-2\operatorname{sn}^2u+k^2\operatorname{sn}^4u}{1-k^2\operatorname{sn}^4u};\\[4pt]\operatorname{dn}2u=\dfrac{\operatorname{dn}^2u-k^2\operatorname{sn}^2u\operatorname{cn}^2u}{1-k^2\operatorname{sn}^4u}=\dfrac{1-2k^2\operatorname{sn}^2u+k^2\operatorname{sn}^4u}{1-k^2\operatorname{sn}^4u}.\end{cases}$$

From these

$$(10)\begin{cases}1-\operatorname{cn}2u=\dfrac{2\operatorname{cn}^2u\operatorname{dn}^2u}{1-k^2\operatorname{sn}^4u};\\[4pt]1+\operatorname{cn}2u=\dfrac{2\operatorname{cn}^2u}{1-k^2\operatorname{sn}^4u};\\[4pt]1-\operatorname{dn}u=\dfrac{2k^2\operatorname{sn}^2u\operatorname{cn}^2u}{1-k^2\operatorname{sn}^4u};\\[4pt]1+\operatorname{dn}u=\dfrac{2\operatorname{dn}^2u}{1-k^2\operatorname{sn}^4u};\end{cases}$$

and therefore

(11)
$$\begin{cases} \operatorname{sn}^2 u = \dfrac{1 - \operatorname{cn} 2u}{1 + \operatorname{dn} 2u}; \\ \operatorname{cn}^2 u = \dfrac{\operatorname{dn} 2u + \operatorname{cn} 2u}{1 + \operatorname{dn} 2u}; \\ \operatorname{dn}^2 u = \dfrac{1 - k^2 + \operatorname{dn} 2u + k^2 \operatorname{cn} 2u}{1 + \operatorname{dn} 2u}; \end{cases}$$

and by analogy

(12)
$$\begin{cases} \operatorname{sn} \dfrac{u}{2} = \sqrt{\dfrac{1 - \operatorname{cn} u}{1 + \operatorname{dn} u}}; \\ \operatorname{cn} \dfrac{u}{2} = \sqrt{\dfrac{\operatorname{cn} u + \operatorname{dn} u}{1 + \operatorname{dn} u}}; \\ \operatorname{dn} \dfrac{u}{2} = \sqrt{\dfrac{1 - k^2 + \operatorname{dn} u + k^2 \operatorname{cn} u}{1 + \operatorname{dn} u}}. \end{cases}$$

In equations (7) making $u = v$, and taking the lower sign, we have

(13)
$$\begin{cases} \operatorname{sn} 0 = 0; \\ \operatorname{cn} 0 = 1; \\ \operatorname{dn} 0 = 1. \end{cases}$$

Likewise, we get by making $u = 0$,

(14)
$$\begin{cases} \operatorname{sn}(-u) = -\operatorname{sn} u; \\ \operatorname{cn}(-u) = +\operatorname{cn} u; \\ \operatorname{dn}(-u) = \operatorname{dn} u. \end{cases}$$

CHAPTER III.

PERIODICITY OF THE FUNCTIONS.

WHEN the elliptic integral

$$\int_0^\phi \frac{d\phi}{\sqrt{1 - k^2 \sin^2 \phi}}$$

has for its amplitude $\frac{\pi}{2}$, it is called, following the notation of Legendre, the *complete* function, and is indicated by K, thus:

$$K = \int_0^{\frac{\pi}{2}} \frac{d\phi}{\sqrt{1 - k^2 \sin^2 \phi}}.$$

When k becomes the complementary modulus, k', (see eq. 4, Chap. IV,) the corresponding complete function is indicated by K', thus:

$$K' = \int_0^{\frac{\pi}{2}} \frac{d\phi}{\sqrt{1 - k'^2 \sin^2 \phi}}.$$

From these, evidently,

$$\text{am}(K, k) = \frac{\pi}{2}, \qquad \text{am}(K', k') = \frac{\pi}{2}.$$

(1)
$$\begin{cases} \text{sn}(K, k) = 1; \\ \text{cn}(K, k) = 0; \\ \text{dn}(K, k) = k'. \end{cases}$$

From eqs. (7), (8), and (9), Chap. II, we have, by the substitution of the values of sn $(K) = 1$, cn $(K) = 0$, dn $(K) = k'$,

(2) $\begin{cases} \text{sn } 2K = 0; \\ \text{cn } 2K = -1; \\ \text{dn } 2K = 1. \end{cases}$

These equations, by means of (1), (2), and (3) of Chap. II, give

(3) $\begin{cases} \text{sn } (u + 2K) = -\text{sn } u; \\ \text{cn } (u + 2K) = -\text{cn } u; \\ \text{dn } (u + 2K) = \text{dn } u; \end{cases}$

and these, by changing u into $u + 2K$, give

(4) $\begin{cases} \text{sn } (u + 4K) = \text{sn } u; \\ \text{cn } (u + 4K) = \text{cn } u; \\ \text{dn } (u + 4K) = \text{dn } u. \end{cases}$

From these equations it is seen that the elliptic functions sn, cn, dn, are periodic functions having for their period $4K$. Unlike the period of trigonometric functions, this period is not a fixed one, but depends upon the value of k, the modulus.

From the Integral Calculus we have

$$\int_0^{n\frac{\pi}{2}} \frac{d\phi}{\Delta\phi} = \int_0^{\frac{\pi}{2}} \frac{d\phi}{\Delta\phi} + \int_{\frac{\pi}{2}}^{\pi} \frac{d\phi}{\Delta\phi} + \int_{\pi}^{\frac{3\pi}{2}} \frac{d\phi}{\Delta\phi} + \cdots + \int_{(n-1)\frac{\pi}{2}}^{n\frac{\pi}{2}} \frac{d\phi}{\Delta\phi}$$

$$= n \int_0^{\frac{\pi}{2}} \frac{d\phi}{\Delta\phi} = nK;$$

from which we see that

$$n\frac{\pi}{2} = \text{am }(nK);$$

or, since $\frac{\pi}{2} = \operatorname{am} K$,

$$\operatorname{am}(nK) = n \cdot \operatorname{am} K,$$

and

$$n\pi = \operatorname{am}(2nK),$$

and also

$$n\pi = 2n \operatorname{am} K.$$

In the case of an Elliptic Integral with the arbitrary angle α, we can put

$$d = n\pi \pm \beta,$$

where β is an angle between 0 and $\frac{\pi}{2}$, the upper or the lower sign being taken according as $\frac{\pi}{2}$ is contained in α an even or an uneven number of times.

In the first case we have

$$\int_0^{n\pi+\beta} \frac{d\phi}{\Delta\phi} = \int_0^{n\pi} \frac{d\phi}{\Delta\phi} + \int_{n\pi}^{n\pi+\beta} \frac{d\phi}{\Delta\phi};$$

or, putting $\phi_1 = \phi - n\pi$,

$$\int_0^{n\pi+\beta} \frac{d\phi}{\Delta\phi} = 2nK + \int_0^{\beta} \frac{d\phi_1}{\Delta\phi_1}.$$

In the second case

$$\int_0^{n\pi-\beta} \frac{d\phi}{\Delta\phi} = \int_0^{n\pi} \frac{d\phi}{\Delta\phi} - \int_{n\pi-\beta}^{n\pi} \frac{d\phi}{\Delta\phi};$$

or, putting $\phi_1 = n\pi - \phi$,

$$\int_0^{n\pi-\beta} \frac{d\phi}{\Delta\phi} = 2nK - \int_0^{\beta} \frac{d\phi_1}{\Delta\phi_1};$$

PERIODICITY OF THE FUNCTIONS. 25

or in either case,

$$\int_0^{n\pi \pm \beta} \frac{d\phi}{\Delta\phi} = 2nK \pm \int_0^{\beta} \frac{d\phi_1}{\Delta\phi_1}.$$

Thus we see that the Integral with the general amplitude α can be made to depend upon the complete integral K and an Integral whose amplitude lies between 0 and $\frac{\pi}{2}$.

Put now

$$\int_0^{\beta} \frac{d\phi_1}{\Delta\phi_1} = u, \qquad \beta = \operatorname{am} u.$$

This gives

$$\int_0^{n\pi \pm \beta} \frac{d\phi}{\Delta\phi} = 2nK \pm u,$$

or $\operatorname{am}(2nK \pm u) = n\pi \pm \beta$
(5) $\qquad\qquad\qquad = n\pi \pm \operatorname{am} u$
(6) $\qquad\qquad\qquad = 2n \cdot \operatorname{am} K \pm \operatorname{am} u;$
or, since $\operatorname{am}(-z) = -\operatorname{am} z,$
$\operatorname{am}(u \pm 2nK) = \operatorname{am} u \pm n\pi$
$\qquad\qquad\qquad = \operatorname{am} u \pm 2n \cdot \operatorname{am} K.$

Taking the sine and cosine of both sides, we have

$$\operatorname{sn}(u + 2nK) = \pm \operatorname{sn} u;$$
$$\operatorname{cn}(u + 2nK) = \pm \operatorname{cn} u;$$

the upper or the lower sign being taken according as n is even or odd. By giving the proper values to n we can get the same results as in equations (3) and (4).

Putting $n = 1$ in eq. (5), we have

$$\operatorname{sn}(2K - u) = \sin \pi \operatorname{cn} u - \cos \pi \operatorname{sn} u$$
(7) $\qquad\qquad\qquad = \operatorname{sn} u.$

Elliptic functions also have an imaginary period. In order to show this we will, in the integral

$$\int_0^\phi \frac{d\phi}{\Delta\phi},$$

assume the amplitude as imaginary. Put

$$\sin\phi = i\tan\psi.$$

From this we get

(8) $$\begin{cases} \cos\phi = \dfrac{1}{\cos\psi}; \\[6pt] \Delta\phi = \dfrac{\sqrt{1-k'^2\sin^2\psi}}{\cos\psi} = \dfrac{\Delta(\psi, k')}{\cos\psi}; \\[6pt] d\phi = i\dfrac{d\psi}{\cos\psi}. \end{cases}$$

From these, since ϕ and ψ vanish simultaneously, we easily get

$$\int_0^\phi \frac{d\phi}{\Delta\phi} = i\int_0^\psi \frac{d\psi}{\Delta(\psi, k')}.$$

Put

$$\int_0^\psi \frac{d\psi}{\Delta(\psi, k')} = u \quad \text{and} \quad \psi = \operatorname{am}(u, k'),$$

whence

$$\int_0^\phi \frac{d\phi}{\Delta\phi} = iu \quad \text{and} \quad \phi = \operatorname{am}(iu);$$

and these substituted in Eq. (8) give

(9) $$\begin{cases} \operatorname{sn} iu = i\operatorname{tn}(u, k'); \\[6pt] \operatorname{cn} iu = \dfrac{1}{\operatorname{cn}(u, k')}; \\[6pt] \operatorname{dn} iu = \dfrac{\operatorname{dn}(u, k')}{\operatorname{cn}(u, k')}. \end{cases}$$

PERIODICITY OF THE FUNCTIONS.

By assuming
$$\int_0^\psi \frac{d\psi}{\Delta(\psi, k')} = iu \quad \text{and} \quad \int_0^\phi \frac{d\phi}{\Delta\phi} = -u,$$
we get
$$\operatorname{sn}(-u) = i \operatorname{tn}(iu, k'),$$
$$\operatorname{cn}(-u) = \frac{1}{\operatorname{cn}(iu, k')},$$
$$\operatorname{dn}(-u) = \frac{\operatorname{dn}(iu, k')}{\operatorname{cn}(iu, k')};$$
or, from eq. (14), Chap. II,

(10)
$$\begin{cases} \operatorname{sn} u = -i \operatorname{tn}(iu, k'); \\ \operatorname{cn} u = \dfrac{1}{\operatorname{cn}(iu, k')}; \\ \operatorname{dn} u = \dfrac{\operatorname{dn}(iu, k')}{\operatorname{cn}(iu, k')}. \end{cases}$$

From eqs. (7), Chap. II, making $v = K$, we get, since $\operatorname{sn} K = 1$, $\operatorname{cn} K = 0$, $\operatorname{dn} K = k'$,

(11)
$$\begin{cases} \operatorname{sn}(u \pm K) = \pm \dfrac{\operatorname{cn} u \operatorname{dn} u}{1 - k^2 \operatorname{sn}^2 u} = \pm \dfrac{\operatorname{cn} u}{\operatorname{dn} u}; \\ \operatorname{cn}(u \pm K) = \dfrac{\mp \operatorname{sn} u \operatorname{dn} u k'}{\operatorname{dn}^2 u} = \mp \dfrac{k' \operatorname{sn} u}{\operatorname{dn} u}; \\ \operatorname{dn}(u \pm K) = + \dfrac{k'}{\operatorname{dn} u}. \end{cases}$$

In these equations, changing u into iu, we get, by means of eqs. (9),

(12)
$$\begin{cases} \operatorname{sn}(iu \pm K) = \pm \dfrac{1}{\operatorname{dn}(u, k')}; \\ \operatorname{cn}(iu \pm K) = \mp \dfrac{ik' \operatorname{sn}(u, k')}{\operatorname{dn}(u, k')}; \\ \operatorname{dn}(iu \pm K) = + \dfrac{k' \operatorname{cn}(u, k')}{\operatorname{dn}(u, k')}. \end{cases}$$

28   ELLIPTIC FUNCTIONS.

Putting now in eqs. (9) $u \pm K'$ instead of u, and making use of eqs. (10), and interchanging k and k', we have

(13) $\begin{cases} \operatorname{sn}(iu \pm iK') = -\dfrac{i \operatorname{cn}(u, k')}{k \operatorname{sn}(u, k')}; \\ \operatorname{cn}(iu \pm iK') = \mp \dfrac{\operatorname{dn}(u, k')}{k \operatorname{sn}(u, k')}; \\ \operatorname{dn}(iu \pm iK') = \mp \dfrac{1}{\operatorname{sn}(u, k')}. \end{cases}$

Substituting in these $-iu$ in place of u, we get, by means of eqs. (9) and eqs. (14) of Chap. II,

(14) $\begin{cases} \operatorname{sn}(u \pm iK') = \dfrac{1}{k \operatorname{sn} u}; \\ \operatorname{cn}(u \pm iK') = \mp \dfrac{i \operatorname{dn} u}{k \operatorname{sn} u}; \\ \operatorname{dn}(u \pm iK') = \mp i \cot \operatorname{am} u. \end{cases}$

In these equations, putting $u + K$ in place of u, we get

(15) $\begin{cases} \operatorname{sn}(u + K \pm iK') = +\dfrac{\operatorname{dn} u}{k \operatorname{cn} u}; \\ \operatorname{cn}(u + K \pm iK') = \mp \dfrac{ik'}{k \operatorname{cn} u}; \\ \operatorname{dn}(u + K \pm iK') = \pm ik' \operatorname{tn} u. \end{cases}$

Whence for $u = 0$ we get

(16) $\begin{cases} \operatorname{sn}(K \pm iK') = \dfrac{1}{k}; \\ \operatorname{cn}(K \pm iK') = \mp \dfrac{ik'}{k}; \\ \operatorname{dn}(K \pm iK') = 0. \end{cases}$

If in eqs. (14) we put $u = 0$, we see that as u approaches zero, the expressions

$$\operatorname{sn}(\pm iK'), \quad \operatorname{cn}(\pm iK'), \quad \operatorname{dn}(\pm iK')$$

approach infinity.

We see from what has preceded that Elliptic Functions have two periods, one a real period, and one an imaginary period.

In the former characteristic they resemble Trigonometric Functions, and in the latter Logarithmic Functions.

On account of these two periods they are often called Doubly Periodic Functions. Some authors make this double periodicity the starting-point of their investigations. This method of investigation gives some very beautiful results and processes, but not of a kind adapted for an elementary work.

It will be noticed that the Elliptic Functions sn u, cn u, and dn u have a very close analogy to trigonometric functions, in which, however, the independent variable u is not an angle, as it is in the case of trigonometric functions.

Like Trigonometric Functions, these Elliptic Functions can be arranged in tables. These tables, however, require a double argument, viz., u and k. In Chap. IX these functions are developed into series, from which their values may be computed and tables formed.

No complete tables have yet been published, though they are in process of computation.

CHAPTER IV.

LANDEN'S TRANSFORMATION.

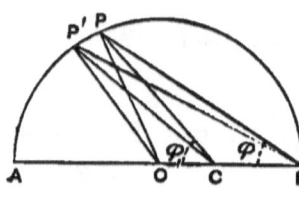

LET AB be the diameter of a circle, with the centre at O, the radius $AO = r$, and C a fixed point situated upon OB, and $OC = k_0 r$. Denote the angle PBC by ϕ, and the angle PCO by ϕ_1. Let P' be a point indefinitely near to P. Then

$$\frac{PP'}{PC} = \frac{\sin PCP'}{\sin PP'C} = \frac{\sin PCP'}{\cos OP'C}.$$

But $PP' = 2rd\phi$, and $\sin PCP' = PCP' = d\phi_1$; therefore

$$\frac{2rd\phi}{PC} = \frac{d\phi_1}{\cos OP'C}.$$

But

$$\overline{PC}^2 = r^2 + r^2 k_0^2 + 2r^2 k_0 \cos 2\phi$$
$$= (r + rk_0)^2 \cos^2 \phi + (r - rk_0)^2 \sin^2 \phi;$$

also

$$r^2 \cos^2 OP'C = r^2 - r^2 \sin^2 OP'C$$
$$= r^2 - r^2 k_0^2 \sin^2 \phi_1.$$

Therefore

$$\frac{2d\phi}{\sqrt{(r + rk_0)^2 \cos^2 \phi + (r - rk_0)^2 \sin^2 \phi}} = \frac{d\phi_1}{\sqrt{r^2 - r^2 k_0^2 \sin^2 \phi_1}},$$

which can be written

$$\frac{2}{r + rk_0} \frac{d\phi}{\sqrt{1 - \frac{4k_0 r^2}{(r + rk_0)^2} \sin^2 \phi}} = \frac{1}{r} \frac{d\phi_1}{\sqrt{1 - k_0^2 \sin^2 \phi_1}},$$

Putting

(1) $$\frac{4k_0 r^2}{(r+rk_0)^2} = \frac{4k_0}{(1+k_0)^2} = k^2,$$

we have

(2) $$\int_0^\phi \frac{d\phi}{\sqrt{1-k^2 \sin^2 \phi}} = \frac{1+k_0}{2} \int_0^{\phi_1} \frac{d\phi_1}{\sqrt{1-k_0^2 \sin^2 \phi_1}};$$

no constant being added because ϕ and ϕ_1 vanish simultaneously; ϕ and ϕ_1 being connected by the equation

(3) $$\frac{\sin OPC}{\sin OCP} = \frac{\sin(2\phi - \phi_1)}{\sin \phi_1} = \frac{rk_0}{r} = k_0.$$

From the value of k^2 we have

(4) $$1 - k^2 = k'^2 = \frac{(1-k_0)^2}{(1+k_0)^2},$$

and therefore

(5) $$k_0 = \frac{1-k'}{1+k'}.$$

k' is called the *complementary modulus*, and is evidently the minimum value of $\Delta\phi$, the value of $\Delta\phi$ when $\phi = 90°$:

$$\sqrt{1-k^2} = k'.$$

From eq. (1) we evidently have $k > k_0$, for, putting eq. (1) in the form

$$\frac{k^2}{k_0^2} = \frac{4}{k_0 + 2k_0^2 + k_0^3},$$

we see that if $k_0 = 1$, then $k = k_0$, but as $k_0 < 1$, always, as is evident from the figure, k must be greater than k_0.

It is also evident, from the figure, that $\phi_1 > \phi$. Or it may be deduced directly from eq. (3).

Since $k < 1$, we can write

$$k = \sin \theta, \qquad k' = \sqrt{1-k^2} = \cos \theta.$$

Substituting in eq. (5), we have

$$k_0 = \frac{1-k'}{1+k'} = \tan^2 \tfrac{1}{2}\theta,$$

and we can write

$$k_0 = \sin \theta_0, \qquad k_1' = \sqrt{1-k_0^2} = \cos \theta_0.$$

From eq. (5) we have

$$1 + k_0 = \frac{2}{1+k'}.$$

Substituting the value of k_0 in that for k_1', we get

$$k_1' = \frac{2\sqrt{k'}}{1+k'}.$$

We also have

$$2\phi - \phi_1 = \phi - (\phi_1 - \phi)$$
$$\phi_1 = \phi + (\phi_1 - \phi),$$

and, eq. (3), becomes

$$\operatorname{sn}(2\phi - \phi_1) = k_0 \sin \phi_1,$$

$$\sin \phi \cos(\phi_1 - \phi) - \cos \phi \sin(\phi_1 - \phi) =$$
$$k_0 \sin \phi \cos(\phi_1 - \phi) + k_0 \cos \phi \sin(\phi_1 - \phi),$$

or

$$\tan \phi - \tan(\phi_1 - \phi) = k_0 \tan \phi + k_0 \tan(\phi_1 - \phi),$$

or

$$\tan(\phi_1 - \phi) = \frac{1-k_0}{1+k_0} \tan \phi$$
$$= k' \tan \phi.$$

Collecting these results, we have

(6) $$k = \frac{2\sqrt{k_0}}{1+k_0} = \sin \theta;$$

(7) $$k_0 = \frac{1-k'}{1+k'} = \sin \theta_0 = \tan^2 \tfrac{1}{2}\theta;$$

(8) $$k_1' = \frac{2\sqrt{k'}}{1+k'} = \cos\theta_0;$$

(9) $$k' = \frac{1-k_0}{1+k_0} = \cos\theta;$$

(10) $$1+k_0 = \frac{2}{1+k'} = \frac{2\sqrt{k_0}}{k} = \frac{k_1'}{\sqrt{k'}} = \frac{1}{\cos^2\tfrac{1}{2}\theta};$$

(11) $$\sin(2\phi - \phi_1) = k_0 \sin\phi_1;$$

(12) $$\tan(\phi_1 - \phi) = k' \tan\phi;$$

(13) $$\int_0^\phi \frac{d\phi}{\Delta(k,\phi)} = \frac{1+k_0}{2} \int_0^{\phi_1} \frac{d\phi_1}{\Delta(k_0,\phi_1)};$$

(14) $$k = \sqrt{1-k'^2},\ k' = \sqrt{1-k^2}.$$

Upon examination it will easily appear that k and k_0, and θ and θ_0, are the first two terms of a decreasing series of moduli and angles; k' and k_1', and ϕ and ϕ_1, of an increasing series; the law connecting the different terms of the series being deduced from eqs. (6) to (12).

By repeated applications of these equations we would get the following series of moduli and amplitudes:

$$k_{0n} = 0_{(n=\infty)} \qquad k_n' = 1_{(n=\infty)} \qquad \phi_n$$
$$\vdots \qquad \vdots \qquad \vdots$$
$$k_{00} \qquad k_2' \qquad \phi_2$$
$$k_0 \qquad k_1' \qquad \phi_1$$
$$k \qquad k' \qquad \phi$$

The upper limit of the one series of moduli is 1, and the lower limit of the other series is 0, as is indicated. k and k', which are bound by the relation $k^2 + k'^2 = 1$, are called the *primitives* of the series.

NOTE.—It will be noticed that the successive terms of a decreasing series are indicated by the sub-accents 0, 00, 03, 04, ... 0n; and the successive terms of an increasing series by the sub-accents 1, 2, 3, ... n.

Again, by application of these equations, we can form a new series running up from k, viz., $k_1, k_2, k_3, \ldots k_n = 1_{(n=\infty)}$; and also a new series running down from k', viz., $k_0', k_{00}', \ldots k_{0n}' = 0_{(n=\infty)}$. So also with ϕ.

Collecting these series, we have

$k_{0n} = 0$	$k_n' = 1$	ϕ_n
.	.	.
.	.	.
.	.	.
k_{02}	k_2'	ϕ_2
k_0	k_1'	ϕ_1
k	k'	ϕ
k_1	k_0'	ϕ_0
k_2	k_{00}'	ϕ_{00}
.	.	.
.	.	.
.	.	.
$k_n = 1$	$k_{0n}' = 0$	$\phi_{0n} = 0$

NOTE.—In practice it will be found that generally n will not need to be very large in order to reach the limiting values of the terms, often only two or three terms being needed.

Applying eqs. (7), (12), (13), and (14) repeatedly, we get

$$(14_1) \begin{cases} k = \sin \theta, & k' = \cos \theta; \\ k_0 = \dfrac{1 - k'}{1 + k'} = \tan^2 \tfrac{1}{2}\theta = \sin \theta_0, & k_1' = \cos \theta_0; \\ k_{00} = \tan^2 \tfrac{1}{2}\theta_0 = \sin \theta_{00}, & k_2' = \cos \theta_{00}; \\ k_{03} = \tan^2 \tfrac{1}{2}\theta_{00} = \sin \theta_{03}, & k_3' = \cos \theta_{03}; \\ \quad \cdots \cdots \cdots \cdots \cdots \cdots \cdots \\ k_{0n} = \tan^2 \tfrac{1}{2}\theta_{0(n-1)} = \sin \theta_{0n}, & k_n' = \cos \theta_{0n}. \end{cases}$$

$$(14_2)\begin{cases} \tan(\phi_1 - \phi) = k' \tan \phi; \\ \tan(\phi_2 - \phi_1) = k_1' \tan \phi_1; \\ \tan(\phi_3 - \phi_2) = k_2' \tan \phi_2; \\ \cdots \cdots \cdots \cdots \\ \tan(\phi_n - \phi_{n-1}) = k'_{(n-1)} \tan \phi_{n-1}. \end{cases}$$

$$(14_3)\begin{cases} F(k, \phi) = \dfrac{1 + k_0}{2} F(k_0, \phi_1); \\[4pt] F(k_0, \phi_1) = \dfrac{1 + k_{00}}{2} F(k_{00}, \phi_2); \\[4pt] F(k_{00}, \phi_2) = \dfrac{1 + k_{03}}{2} F(k_{03}, \phi_3); \\ \cdots \cdots \cdots \cdots \\ F(k_{0(n-1)}, \phi_{n-1}) = \dfrac{1 + k_{0n}}{2} F(k_{0n}, \phi_n). \end{cases}$$

Multiplying these latter equations together, member by member, we have

$$(15) \quad F(k, \phi) = (1 + k_0)(1 + k_{00}) \cdots (1 + k_{0n}) \frac{F(k_{0n}, \phi_n)}{2^n};$$

k_0, k_{00}, etc., and ϕ_1, ϕ_2, etc., being determined from the preceding equations.

From eqs. (9) and (10) we get

$$1 + k_0 = \frac{1}{\cos^2 \tfrac{1}{2}\theta}, \quad 1 + k_{00} = \frac{1}{\cos^2 \tfrac{1}{2}\theta_0}, \quad \text{etc.}$$

Substituting these in eq. (15), we get

$$(16) \quad F(k, \phi) = \frac{1}{\cos^2 \dfrac{\theta}{2} \cos^2 \dfrac{\theta_0}{2} \cdots \cos^2 \dfrac{\theta_{0n}}{2}} \cdot \frac{F(k_{0n}, \phi_n)}{2^n}.$$

From eqs. (15) and (10) we get

$$F(k, \phi) = \sqrt{\frac{k_1' k_2' k_3' \ldots k_n'^2}{k'}} \cdot \frac{F(k_{0n}, \phi_n)}{2^n}.$$

And this with equations (8) and (9) gives

(17) $$F(k, \phi) = \sqrt{\frac{\cos \theta_0 \cos \theta_{00} \ldots \cos^2 \theta_{0n}}{\cos \theta}} \cdot \frac{F(k_{0n}, \phi_n)}{2^n}.$$

Applying equation (13) to (k_1, ϕ_0), (k_2, ϕ_{00}), etc., we get

$$F(k_1, \phi_0) = \frac{1+k}{2} F(k, \phi), \text{ etc.};$$

but since, eq. (10),

$$\frac{1+k}{2} = \frac{1}{1+k_0'}, \text{ etc.,}$$

these become

$$F(k, \phi) = (1 + k_0') F(k_1, \phi_0);$$
$$F(k_1, \phi_0) = (1 + k_{00}') F(k_2, \phi_{00});$$
$$\cdot \cdot \cdot \cdot \cdot \cdot \cdot \cdot$$
$$F(k_{n-1}, \phi_{0(n-1)}) = (1 + k_{0n}') F(k_n, \phi_{0n});$$

whence

(18) $$F(k, \phi) = (1 + k_0')(1 + k_{00}') \ldots (1 + k_{0n}') F(k_n, \phi_{0n}),$$

in which k_0', k_{00}', etc., k_1, k_2, etc., ϕ_0, ϕ_{00}, etc., are determined as follows:

Let
$$k = \sin \theta,$$
$$k_1 = \sin \theta_1.$$

From eq. (10),

$$k_1 = \frac{2\sqrt{k}}{1+k} \quad \text{or} \quad \sin \theta_1 = \frac{2\sqrt{\sin \theta}}{1 + \sin \theta}.$$

Solving this equation for sin θ, we get

$$\sin \theta = \tan^2 \tfrac{1}{2}\theta_1.$$

Hence we can write

(18_1)
$$\begin{cases} k = \sin \theta = \tan^2 \tfrac{1}{2}\theta_1; \\ k_1 = \sin \theta_1 = \tan^2 \tfrac{1}{2}\theta_2; \\ \cdot\ \cdot\ \cdot\ \cdot\ \cdot\ \cdot\ \cdot \\ k_n = \sin \theta_n. \end{cases}$$

From equation (12) we get

(18_2)
$$\begin{cases} \sin (2\phi_0 - \phi) = k \sin \phi;^* \\ \sin (2\phi_{00} - \phi_0) = k_1 \sin \phi_0; \\ \cdot\ \cdot\ \cdot\ \cdot\ \cdot\ \cdot\ \cdot\ \cdot \\ \sin (2\phi_{0n} - \phi_{0(n-1)}) = k_{n-1} \sin \phi_{0(n-1)}. \end{cases}$$

---

* When $\sin \phi = 1$ nearly, ϕ is best determined as follows: From eq. (12) we have

$$\tan (\phi - \phi_0) = k_0' \tan \phi_0$$
$$= k_0' \tan \phi \text{ nearly};$$

whence

$$\phi - \phi_0 \doteq R k_0' \tan \phi \text{ nearly},$$

R being the radian in seconds, viz. 206264".806, and log $R = 5.3144251$. Substituting the approximate value of ϕ_0, we can get a new approximation.

Example. $\phi_0 = 82° 30'$ $k'_{00} = \log^{-1} 5.8757219$

tan 82° 30'	10.8805709
k'_{00}	5.8757219
R	5.3144251
	2.0707179 117".684 = 1'.9614

$\phi_0 - \phi_{00} = 1'.9614$

$\phi_{00} = 82° 28'.0386$ 1st approximation.

To determine k_0', k_{00}', etc., we have

$$(18_1) \begin{cases} k' = \sin \eta, & k = \cos \eta; \\ k_0' = \dfrac{1-k}{1+k} = \tan^2 \tfrac{1}{2}\eta = \sin \eta_0, & k_1 = \cos \eta_0; \\ k_{00}' = \tan^2 \tfrac{1}{2}\eta_0 = \sin \eta_{00}, & k_2 = \cos \eta_{00}; \\ \quad\text{etc.} \qquad\qquad \text{etc.} & \text{etc.} \end{cases}$$

Or, since $1 + k_0' = \dfrac{1}{\cos^2 \tfrac{1}{2}\eta}$, $1 + k_{00}' = \dfrac{1}{\cos^2 \tfrac{1}{2}\eta_0}$, etc.,

we can put eq. (18) in the following form:

$$(19) \quad F(k, \phi) = \dfrac{1}{\cos^2 \tfrac{1}{2}\eta \cos^2 \tfrac{1}{2}\eta_0 \ldots \cos^2 \tfrac{1}{2}\eta_{0n}} F(k_n, \phi_{0n}).$$

From equation (13) we have

$$(19)^* \quad F(k_1, \phi_0) = \dfrac{1+k}{2} F(k, \phi),$$

whence

$$F(k, \phi) = \dfrac{2}{1+k} F(k_1, \phi_0).$$

By repeated applications this gives, after combining,

$$F(k, \phi) = \dfrac{2}{1+k} \cdot \dfrac{2}{1+k_1} \cdots \dfrac{2}{1+k_{n-1}} \cdot F(k_n, \phi_{0n})$$

This value gives

$\phi_0 - \phi_{00} = 117''.1675 = 1'.95279$

$\therefore \phi_{00} = 82° 28'.04721$ 2d approximation.

This value gives

$\phi_0 - \phi_{00} = 117''.1698 = 1'.95283$

$\phi_{00} = 82° 28'.04717$ 3d approximation.

$$= \frac{k_1}{\sqrt{k}} \cdot \frac{k_2}{\sqrt{k_1}} \cdots \frac{k_n}{\sqrt{k_{n-1}}} \cdot F(k_n, \phi_{0n});$$

(20) $\quad F(k, \phi) = \sqrt{\dfrac{k_1 k_2 \cdots k_n^2}{k}} \cdot F(k_n, \phi_{0n});$

k_1, k_2, etc., being determined by repeated applications of

$$k_1 = \frac{2\sqrt{k}}{1+k},$$

or by equations (18$_1$).

In equation (19)* let us change k_1 and ϕ_0 into k' and ϕ respectively, so that the first member may have for its complete function

$$K' = F(k', \phi).$$

Upon examination of eq. (19)* we see that the modulus in the second member must be the one next less than the one in the first member, that is, k_0'; and likewise that the amplitude must be the one next greater than the amplitude in the first member, viz., ϕ_1; hence we get

$$F(k', \phi) = \frac{1 + k_0'}{2} F(k_0', \phi_1).$$

Indicating the complete functions by K' and K_0', we have, since $\phi = \dfrac{\pi}{2}$ when $\phi_1 = \pi$ (see Chap. V),

$$K' = (1 + k_0')K_0';$$

and in the same manner,

$$K_0' = (1 + k'_{00})K'_{00},$$
$$K'_{00} = (1 + k'_{03})K'_{03},$$
$$\cdots \cdots \cdots$$
$$K'_{0(n-1)} = (1 + k'_{0n})K'_{0n};$$

whence
$$K' = (1+k_0')(1+k_{00}')\cdots(1+k_{on}')K'_{on}.$$
Since
$$K'_{on} = \int_0^{\frac{\pi}{2}} d\phi = \frac{\pi}{2}, \qquad (n = \text{limit,})$$
we have

(20)* $\quad (1+k_0')(1+k_{00}')\cdots(1+k_{on}') = \dfrac{2K'}{\pi}.$

From eq. (19)* we have, since [eq. (10), Chap IV]
$$\frac{1+k}{2} = \frac{1}{1+k_0'},$$
$$(1+k_0')\int_0^{\phi_0} \frac{d\phi_0}{\Delta(\phi_0, k_1)} = \int_0^{\phi} \frac{d\phi}{\Delta(\phi_1, k)};$$
whence also, since for $\phi_0 = \dfrac{\pi}{2}$, $\phi = \pi$,
$$(1+k_0') K_1 = 2K,$$
$$(1+k_{00}') K_2 = 2K_1,$$
$$\cdots\cdots\cdots\cdots$$
$$(1+k_{on}') K_n = 2K_{n-1},$$
and
$$(1+k_0')(1+k_{00}')\cdots(1+k_{on}')K_n = 2^n K;$$
or
$$\frac{K_n}{2^n} = \frac{K}{(1+k_0')(1+k_{00}')\cdots} \qquad (n=\infty)$$

(21) $\qquad\qquad = \dfrac{\pi}{2K_1} K.$

Let us find the limiting value of $F(k_{on}, \phi_n)$ in eq. (15). In the equation $\tan(\phi_n - \phi_{n-1}) = k_{n-1} \tan \phi_{n-1}$, we see that when k_{n-1} reaches the limit 1, then $\phi_n - \phi_{n-1} = \phi_{n-1}$ or $\phi_n = 2\phi_{n-1}$. Therefore

$$\frac{\phi_n}{2^n} = \frac{2\phi_{n-1}}{2^n} = \frac{\phi_{n-1}}{2^{n-1}};$$

$$\frac{\phi_{n+1}}{2^{n+1}} = \frac{2\phi_n}{2^{n+1}} = \frac{\phi_n}{2^n} = \frac{\phi_{n-1}}{2^n};$$

$$\frac{\phi_{n+m}}{2^{n+m}} = \frac{\phi_{n-1}}{2^n} = \text{constant, whatever } m \text{ may be.}$$

Therefore eq. (15) becomes

(21)* $\qquad F(k, \phi) = (1 + k_0)(1 + k_{00}) \cdots (1 + k_{0n}) \dfrac{\phi_n}{2^n}$,

n being whatever number will carry k_0 and $\dfrac{\phi_1}{2}$ to their limiting values.

In the same way, eqs. (16) and (17) become

(22) $\qquad F(k, \phi) = \dfrac{1}{\cos^2 \dfrac{\theta}{2} \cos^2 \dfrac{\theta_0}{2} \cdots \cos^2 \dfrac{\theta_{0n}}{2}} \cdot \dfrac{\phi_n}{2^n}$

(23) $\qquad\qquad = \sqrt{\dfrac{\cos \theta_0 \cos \theta_{00} \cdots \cos^2 \theta_{0n}}{\cos \theta}} \cdot \dfrac{\phi_n}{2^n}$,

$n - 1$ being the number which makes $k'_{n-1} = 1$.

In these last three equation k_0, k_{00} are determined by eqs. (14_1); ϕ_1, ϕ_2, etc., by eqs. (14_2)*; θ, θ_0, etc., by eqs. (14_4); and k', k'_1, k'_2, etc., for use in eq. (14_2) by eqs. (14_1).

---

* Taking for $\phi_1 - \phi$, etc., not always the least angle given by the tables, but that which is nearest to ϕ.

BISECTED AMPLITUDES.

We have identically

$$u = 2 \cdot \frac{u}{2} = 2 \int \frac{d\,\text{am}\,\frac{u}{2}}{\sqrt{1 - k^2 \text{sn}^2 \frac{u}{2}}};$$

$$\frac{u}{2} = 2 \cdot \frac{u}{4} = 2F\left(k,\,\text{am}\,\frac{u}{4}\right);$$

etc.

Therefore

$$u = F(k,\,\text{am}\,u) = 2^n F\left(k,\,\text{am}\,\frac{u}{n}\right)$$

$$= 2^n \cdot \text{am}\frac{u}{n}, \qquad (n = \text{limit,})$$

$\text{am}\,\dfrac{u}{n}$ being determined by repeated applications of eq. (12) of Chap. II, as follows:

$$\text{sn}^2 \frac{u}{2} = \frac{1 - \text{cn}\,u}{1 + \text{dn}\,u} = \frac{2 \sin^2 \frac{1}{2}\,\text{am}\,u}{1 + \text{dn}\,u};$$

(24) $$\text{sn}\,\frac{u}{2} = \frac{\sin \frac{1}{2}\,\text{am}\,u}{\sqrt{\dfrac{1 + \cos \beta}{2}}} = \frac{\sin \dfrac{\text{am}\,u}{2}}{\cos \frac{1}{2}\beta};$$

β being an angle determined by the equation

(25) $$\cos \beta = \text{dn}\,u = \sqrt{1 - k^2 \text{sn}^2 u},$$

and n being the number which makes

$$2^n\,\text{am}\,\frac{u}{n} = \text{constant}.$$

$\text{am}\,\dfrac{u}{n}$ is found by repeated applications of eq. (24).

Indicating the amplitudes as follows:

$$\text{am } u = \phi,$$

$$\text{am } \frac{u}{2} = \phi_{02},$$

$$\text{am } \frac{u}{4} = \phi_{04},$$

$$\text{am } \frac{u}{8} = \phi_{08},$$

$$\cdots \cdots$$

$$\text{am } \frac{u}{2^n} = \phi_{02^n}, —$$

(26) $$F(k, \phi) = 2^n \phi_{02^n};$$

n being the limiting value.

In eq. (18), when k_n reaches its limit 1, we have

$$F(k_n, \phi_{on}) = \int_0^\phi \frac{d\phi_{on}}{\cos \phi_{on}} = \log_e \tan(45° + \tfrac{1}{2}\phi_{on}),$$

and eqs. (18) and (19) become

(27) $$F(k, \phi) = (1 + k_0')(1 + k_{00}') \cdots (1 + k_{on}') \log_e \tan(45° + \tfrac{1}{2}\phi_{on})$$

$$= \frac{1}{\cos^2 \tfrac{1}{2}\eta \cos^2 \tfrac{1}{2}\eta_0 \cdots \cos^2 \tfrac{1}{2}\eta_{on}} \log_e \tan(45° + \tfrac{1}{2}\phi_{on})$$

(28) $$= \frac{1}{\cos^2 \tfrac{1}{2}\eta \cos^2 \tfrac{1}{2}\eta_0 \cdots \cos^2 \tfrac{1}{2}\eta_{on}} \cdot \frac{1}{M} \log \tan(45° + \tfrac{1}{2}\phi_{on});$$

n being the number which renders $k_n = 1$.

Eq. (20) becomes

(29) $$F(k, \phi) = \sqrt{\frac{k_1 k_2 \ldots k_n^2}{k}} \cdot \log_e \tan(45° + \tfrac{1}{2}\phi_{on})$$

$$= \sqrt{\frac{k_1 k_2 \ldots k_n^2}{k}} \cdot \frac{1}{M} \log \tan(45° + \tfrac{1}{2}\phi_{on})$$

$$= \sqrt{\frac{\cos \eta_0 \cos \eta_{00} \ldots \cos^2 \eta_{on}}{\cos \eta}} \cdot \frac{1}{M} \log \tan(45° + \tfrac{1}{2}\phi_{on}).$$

In these equations k_0', k_{00}', etc., are determined by eqs. (18$_3$); η, η_0, etc., by eqs. (18$_3$); ϕ_0, ϕ_{00}, etc., by eqs. (18$_2$); k_1, k_2, etc., by eqs. (18$_1$).

Substituting in eq. (27) from eq. (20)*, we have

$$F(k, \phi) = \frac{2K'}{\pi} \log_e \tan(45° + \tfrac{1}{2}\phi_{on})$$

(30)
$$= \frac{2K'}{\pi M} \log \tan(45° + \tfrac{1}{2}\phi_{on}).$$

CHAPTER V.

COMPLETE FUNCTIONS.

INDICATE by K the complete integral

(1) $$K = \int_0^{\frac{\pi}{2}} \frac{d\phi}{\sqrt{1 - k^2 \sin^2 \phi}},$$

and by K_0 the complete integral

(2) $$K_0 = \int_0^{\frac{\pi}{2}} \frac{d\phi_1}{\sqrt{1 - k_0^2 \sin^2 \phi_1}};$$

and in a similar manner K_{00}, K_{03}, etc.

From eq. (12), Chap. IV, we have

$$\tan(\phi_1 - \phi) = k' \tan \phi$$

$$= \frac{\tan \phi_1 - \tan \phi}{1 + \tan \phi_1 \tan \phi},$$

whence

$$\tan \phi_1 = \frac{(1 + k') \tan \phi}{1 - k' \tan^2 \phi}$$

$$= \frac{1 + k'}{\dfrac{1}{\tan \phi} - k' \tan \phi}.$$

From this equation we see that when $\phi = \frac{\pi}{2}$, $\phi_1 = \pi$. This same result might also have been deduced from Fig. 1, Chap. IV, or from the equation

(3) $\quad \phi_1 = 2\phi - k_0 \sin 2\phi + \tfrac{1}{2}k_0^2 \sin 4\phi - \text{etc.},$

this last being the well-known trigonometrical formula

$$\tan x = n \tan y,$$

$$x = y - \frac{1-n}{1+n}\sin 2y + \frac{1}{2}\left(\frac{1-n}{1+n}\right)^2 \sin 4y - \frac{1}{3}\left(\frac{1-n}{1+n}\right)^3 \sin 6y + \text{etc.}$$

Since $\displaystyle\int_0^{\frac{\pi}{2}} \frac{d\phi_1}{\Delta(k_0\phi_1)} = K_0$, we must have

$$\int_0^{\pi} \frac{d\phi_1}{\Delta(k_0\phi_1)} = 2K_0.$$

These values substituted in eq. (13), Chap. IV, give successively

(4) $\quad K = (1 + k_0)K_0,$
$\quad\quad K_0 = (1 + k_{00})K_{00},$
$\quad\quad \cdot\ \cdot\ \cdot\ \cdot\ \cdot\ \cdot$
$\quad\quad K_{0(n-1)} = (1 + k_{0n})K_{0n};$

whence

(5) $\quad K = (1+k_0)(1+k_{00}) \ldots (1+k_{0n})K_{0n}.$

Since the limit of k_{0n} is 0, K_{0n} becomes

$$K_{0n} = \int_0^{\frac{\pi}{2}} d\phi = \frac{\pi}{2},$$

and we have

(6) $$K = \frac{\pi}{2}(1 + k_0)(1 + k_{00}) \ldots$$

(7) $$= \frac{\frac{1}{2}\pi}{\cos^2 \frac{1}{2}\theta \cos^2 \frac{1}{2}\theta_0 \ldots \cos^2 \frac{1}{2}\theta_{0n}};$$

k_1, k_0, etc., and θ_1, θ_0, etc., being found by eqs. (14$_1$) of Chap. IV.

From the formulæ in these two chapters we can compute the values of u for all values of ϕ and k and arrange them in tables. These are Legendre's Tables of Elliptic Integrals.

CHAPTER VI.

EVALUATION FOR ϕ.

TO FIND ϕ, u AND k BEING GIVEN.

From eqs. (21) and (23), Chap. IV, we have (n having the value which makes $\cos \theta_{0n} = 1$)

$$(1) \quad \phi_n = \frac{2^n u}{(1 + k_0)(1 + k_{00}) \cdots (1 + k_{0n})} = \frac{2^n u \sqrt{\cos \theta}}{\sqrt{\cos \theta_0 \cdots \cos^2 \theta_{0n}}},$$

from which ϕ_n can be calculated, k_0, k_{00}, etc., being found by means of equations (14,), Chap. IV.

Then, having ϕ_n, k_0, k_{00}, etc., we can find ϕ by means of the following equations:

$$\sin(2\phi_{n-1} - \phi_n) = k_{0n} \sin \phi_n,$$
$$\sin(2\phi_{n-2} - \phi_{n-1}) = k_{0(n-1)} \sin \phi_{n-1},$$
$$\cdots \cdots \cdots \cdots$$
$$\sin(2\phi - \phi_1) = k_0 \sin \phi_1;$$

whence we can get the angle ϕ.

When $k > \sqrt{\frac{1}{2}}$ the following formulæ will generally be found to work more rapidly:

From eq. (29), Chap. IV, we have

$$(2) \quad \log \tan(45° + \tfrac{1}{2}\phi_{0n}) = \frac{uM}{\sqrt{\dfrac{k_1 k_2 \cdots k_n^2}{k}}},$$

from which we can get ϕ_{on}; k_1, k_2, etc., being calculated from eqs. (18,), Chap. IV, and ϕ being calculated from the following equations:

$$\tan(\phi_{o(n-1)} - \phi_{on}) = k_n \tan \phi_{on},$$

$$\cdot \cdot \cdot \cdot \cdot \cdot \cdot \cdot \cdot \cdot$$

$$\tan(\phi_o - \phi_{oo}) = k_2 \tan \phi_{o2},$$

$$\tan(\phi - \phi_\infty) = k \tan \phi_o;$$

whence we get ϕ.

This gives a method of solving the equation

$$F\psi = nF\phi,$$

where n and ϕ and the moduli are known, and ψ is the required quantity. n and ϕ give $F\psi$, and then ψ can be determined by the foregoing methods.

When $k = 1$ *nearly*, equation (2) takes a special form,—

1°. When $\tan \phi$ is very much less than $\dfrac{1}{k'}$. In this case

$$F(k, \phi) = \int \frac{d\phi}{\sqrt{\cos^2 \phi + k'^2 \sin^2 \phi}} = \int \frac{d\phi}{\sqrt{(1 + k'^2 \tan^2 \phi)}\cos^2 \phi}$$

$$= \int \frac{d\phi}{\cos \phi} = \log \tan(45° + \tfrac{1}{2}\phi);$$

whence we can find ϕ.

2°. When $\tan \phi$ and $\dfrac{1}{k'}$ approach somewhat the same value, and $k' \tan \phi$ cannot be neglected, $F(k, \phi)$ must be transposed into another where k' shall be much smaller, so that $k' \tan \phi$ can be neglected.

These methods for finding ϕ apply only when $\phi < \dfrac{\pi}{2}$, that is, $u < K$. In the opposite case ($u > K$) put

$$u = 2nK \pm \nu,$$

the upper or the lower sign being taken according as K is continued in u an even or an odd number of times. In either case $\nu < K$, and we can find ν by the preceding methods.

Having found ν, we have from eq. (5), Chap. III,

$$\operatorname{am} u = \operatorname{am}(2nK \pm \nu)$$
$$= n\pi \pm \operatorname{am} \nu.$$

CHAPTER VII.

DEVELOPMENT OF ELLIPTIC FUNCTIONS INTO FACTORS.

FROM eq. (12), Chap. IV, we readily get

$$\sin(2\phi_0 - \phi) = k \sin \phi;$$

$$\sin \phi = \frac{\sin 2\phi_0}{\sqrt{1 + k^2 + 2k \cos 2\phi_0}}$$

$$= \frac{\sin 2\phi_0}{\sqrt{(1 + k^2) - 4k \sin^2 \phi_0}}$$

$$= \frac{1 + k_0'}{2} \cdot \frac{\sin 2\phi_0}{\sqrt{1 - k_1^2 \sin^2 \phi_0}}$$

$\left(\text{since } \frac{4k}{1+k} = k_1, \text{ and } 1+k = \frac{2}{1+k_0'}, \text{ eqs. (6) and (10), Chap. IV}\right);$
and thence

(1) $$\sin \phi = \frac{(1 + k_0') \sin \phi_0 \cos \phi_0}{\Delta(\phi_0, k_1)}.$$

From eq. (13), Chap. IV, we have

$$\int_0^{\phi_0} \frac{d\phi_0}{\Delta(\phi_0, k_1)} = \frac{1 + k}{2} \int_0^{\phi} \frac{d\phi}{\Delta(\phi, k)};$$

and from eq. (4), Chap. V, passing up the scale of moduli one step,

$$1 + k = \frac{K_1}{K},$$

whence
$$F(\phi_0, k_1) = \frac{K_1}{2K} F(\phi, k).$$

Put
$$F(\phi_0, k_1) = u_1 \quad \text{and} \quad F(\phi, k) = u,$$

whence
$$u_1 = \frac{K_1}{2K} u.$$

Furthermore,
$$\phi = \text{am}(u, k);$$

$$\phi_1 = \text{am}(u_1, k_1) = \text{am}\left(\frac{K_1}{2K} u, k_1\right).$$

Substituting these values in eq. (1), we have

$$\text{sn}(u, k) = (1 + k_0') \frac{\text{sn}\left(\frac{K_1}{2K} u, k_1\right) \text{cn}\left(\frac{K_1}{2K} u, k_1\right)}{\text{dn}\left(\frac{K_1}{2K} u, k_1\right)}.$$

But from eq. (11), Chap. III, we have

$$\frac{\text{cn}(\nu, k_1)}{\text{dn}(\nu, k_1)} = \text{sn}(\nu + K_1, k_1),$$

or

$$\frac{\text{cn}\left(\frac{K_1}{2K} u, k_1\right)}{\text{dn}\left(\frac{K_1}{2K} u, k_1\right)} = \text{sn}\left(\frac{K_1 u}{2K} + K_1, k_1\right)$$

$$= \text{sn}\left(\frac{K_1}{2K}(u + K_1), k_1\right);$$

DEVELOPMENT INTO FACTORS. 53

whence

(2) $\quad \operatorname{sn}(u, k) = (1 + k_0') \operatorname{sn} \dfrac{K_1 u}{2K} \operatorname{sn}\left[\dfrac{K_1}{2K}(u + 2K)\right].$†

(Mod. $= k_1$.)

From this equation, evidently, we have generally

(2)* $\quad \operatorname{sn}(\nu, k_n) = (1 + k_{0\,(n+1)}') \operatorname{sn} \dfrac{K_{n+1}}{2K_n} \nu \operatorname{sn}\left[\dfrac{K_{n+1}}{2K_n}(\nu + 2K_n)\right].$

(Mod. $= k_{n+1}$.)

Applying this general formula to the two factors of eq. (2), we have

$$\operatorname{sn}\left(\dfrac{K_1 u}{2K}, k_1\right) = (1 + k_{00}') \operatorname{sn} \dfrac{K_2}{2K_1} \cdot \dfrac{K_1 u}{2K} \cdot \operatorname{sn}\left[\dfrac{K_2}{2K_1}\left(\dfrac{K_1 u}{2K} + 2K_1\right)\right]$$

(Mod. k_2)

$$= (1 + k_{00}') \operatorname{sn} \dfrac{K_2 u}{2^2 K} \operatorname{sn} \dfrac{K_2}{2^2 K}(u + 4K); \quad (\text{Mod. } k_2;)$$

(3) $\quad \operatorname{sn}\left[\dfrac{K_1}{2K}(u + 2K), k_1\right] = (1 + k_{00}') \operatorname{sn}\dfrac{K_2}{2^2 K}(u + 2K)$

$\quad \cdot \operatorname{sn} \dfrac{K_2}{2K_1}\left[\dfrac{K_1}{2K}(u + 2K) + 2K_1\right].$ (Mod. k_2.)

The last argument in this equation is equal to

$$\dfrac{K_2}{2^2 K}(u + 6K);$$

and since, eq. (7), Chap. III,

$$\operatorname{sn}(u, k_2) = \operatorname{sn}(2K_2 - u, k_2),$$

---

† The analogous formula in Trigonometry is

$$\sin \phi = \tfrac{1}{4} \sin \tfrac{1}{2}\phi \sin \tfrac{1}{2}(\phi + \pi).$$

54   ELLIPTIC FUNCTIONS.

we can put in place of this,

$$2K_2 - \frac{K_2}{2^2K}(u+6K) = \frac{K_2}{2^2K}(2K-u);$$

whence eq. (3) becomes

$$\operatorname{sn}\left[\frac{K_1}{2K}(u+2K), k_1\right] = (1+k'_{00})\operatorname{sn}\frac{K_2}{2^2K}(2K+u)$$

$$\cdot \operatorname{sn}\frac{K_2}{2^2K}(2K-u). \qquad (\text{Mod. } k_2.)$$

Substituting these values in eq. (2), we have

(4) $\quad \operatorname{sn}(u, k) = (1+k_0)(1+k'_{00})^2 \operatorname{sn}\frac{K_2 u}{2^2K}$

$$\cdot \operatorname{sn}\frac{K_2}{2^2K}(2K \pm u) \operatorname{sn}\frac{K_2}{2^2K}(4K+u), \quad (\text{Mod. } k_2,)$$

in which the double sign indicates two separate factors which are to be multiplied together.

By the application of the general equation (2)* we find that the arguments in the second member of eq. (4) will each give rise to two new arguments, as follows:

$$\frac{K_2 u}{2^2 K} \quad \text{gives} \quad \frac{K_2 u}{2^2 K},$$

and

$$\frac{K_2}{2K_2}\left(\frac{K_2 u}{2^2K} + 2K_2\right) = \frac{K_2}{2^2K}(u+8K);$$

$$\frac{K_2}{2^2K}(2K \pm u) \quad \text{gives} \quad \frac{K_2}{2^2K}(2K \pm u),$$

DEVELOPMENT INTO FACTORS. 55

and

$$\frac{K_s}{2K_s}\left[\frac{K_s}{2^sK}(2K\pm u)+2K_s\right]=\frac{K_s}{2^sK}(\mathrm{10}K\pm u),\quad\ldots\ (a)$$

$$\frac{K_s}{2^sK}(4K+u)\quad\text{gives}\quad\frac{K_s}{2^sK}(4K+u),$$

and

$$\frac{K_s}{2K_s}\left[\frac{K_s}{2^sK}(4K+u)+2K_s\right]=\frac{K_s}{2^sK}(\mathrm{12}K+u).\quad\ldots\ (b)$$

Subtracting (a) and (b) from $2K_s$, by which the sine of the amplitudes will not be changed [eq. (7), Chap. III], and since our new modulus is k_s, we have for the expressions (a) and (b),

$$\frac{K_s}{2^sK}(6K\mp u);\quad\ldots\ldots\ldots\ (a')$$

$$\frac{K_s}{2^sK}(4K-u).\quad\ldots\ldots\ldots\ (b')$$

Substituting these values in eq. (4), and remembering the factor $(1+k'_{0s})$ introduced by each application of eq. (2)*, we have

$$\operatorname{sn}(u,k)=(1+k_0')(1+k'_{00})^2(1+k'_{0s})^4\operatorname{sn}\frac{K_s u}{2^sK};$$

$$\cdot\operatorname{sn}\frac{K_s}{2^sK}(2K\pm u)\operatorname{sn}\frac{K_s}{2^sK}(4K\pm u);$$

$$\cdot\operatorname{sn}\frac{K_s}{2^sK}(6K\pm u)\operatorname{sn}\frac{K_s}{2^sK}(8K+u).\quad(\text{Mod. }k_s.)$$

From this the law governing the arguments is clear, and we can write for the general equation

$$\text{(5)} \quad \operatorname{sn}(u, k) = (1 + k_0')(1 + k_{00}')^2(1 + k_{03}')^4 \cdots (1 + k_{0n}')^{2^{n-1}}$$

$$\cdot \operatorname{sn} \frac{K_n u}{2^n K} \operatorname{sn} \frac{K_n}{2^n K} (2K \pm u)$$

$$\cdot \operatorname{sn} \frac{K_n}{2^n K} (4K \pm u) \operatorname{sn} \frac{K_n}{2^n K} (6K \pm u)$$

$$\cdots \operatorname{sn} \frac{K_n}{2^n K} \left[(2^n - 2)K \pm u\right]$$

$$\cdot \operatorname{sn} \frac{K_n}{2^n K} (2^n K + u). \qquad \text{(Mod. } k_n.\text{)}$$

Indicate the continued product of the binomial factors by A', and we have

$$A' = (1 + k_0')(1 + k_{00}')^2(1 + k_{03}')^4(1 + k_{04}')^8 \cdots$$

Since the limit of k_0', k_{00}', etc., is zero, it is evident that these factors converge toward the value unity. It can be shown that the functional factors also converge toward the value unity. Thus the argument of the last factor can be written

$$K_n + \frac{K_n u}{2^n K}.$$

From eq. (11), Chap. III, we get then

$$\operatorname{sn}\left(K_n + \frac{K_n u}{2^n K}\right) = \frac{\operatorname{cn} \dfrac{K_n u}{2^n K}}{\operatorname{dn} \dfrac{K_n u}{2^n K}}. \qquad \text{(Mod. } k_n.\text{)}$$

But since k_n at its limit is equal to unity, $\operatorname{cn} = \operatorname{dn}$; whence the last factor of eq. (5) is unity.

DEVELOPMENT INTO FACTORS.

From eq. (21), Chap. IV, we have

$$\lim \frac{K_n}{2^n K} = \frac{2\pi}{2K'}.$$

Therefore for $n = \infty$, eq. (5) becomes

$$\operatorname{sn}(u, k) = A' \operatorname{sn} \frac{\pi u}{2K'} \operatorname{sn} \frac{\pi}{2K'} (2K \pm u)$$

$$\cdot \operatorname{sn} \frac{\pi}{2K'} (4K \pm u) \operatorname{sn} \frac{\pi}{2K'} (6K \pm u), \ \ldots \ .$$

(Mod. 1,)

or

(6) $\qquad \operatorname{sn}(u, k) = A' \operatorname{sn} \frac{\pi u}{2K'} \left[\overset{\infty}{\underset{1}{\Pi^h}} \right] \operatorname{sn} \frac{\pi}{2K'} (2hK \pm u),$

(Mod. 1,)

where the sign $[\Pi]$ indicates the continued product in the same manner as Σ indicates the continued sum.

When $k = 1$, $\int_0^\phi F(\phi, k)$ becomes

$$v = \int_0^\phi \frac{d\phi}{\cos \phi} = \tfrac{1}{2} \log \cdot \frac{1 + \sin \phi}{1 - \sin \phi};$$

whence

$$e^{2v} = \frac{1 + \sin \phi}{1 - \sin \phi},$$

and

$$\sin \phi = \frac{e^{2v} - 1}{e^{2v} + 1} = \frac{e^v - e^{-v}}{e^v + e^{-v}} = \operatorname{sn}(v, 1).$$

Hence in equation (6)

$$\sin \frac{\pi u}{2K'} = \frac{e^{\frac{\pi u}{2K'}} - e^{-\frac{\pi u}{2K'}}}{e^{\frac{\pi u}{2K'}} + e^{-\frac{\pi u}{2K'}}};$$

$$\operatorname{sn} \frac{\pi}{2K'}(2hK \pm u) = \frac{e^{\frac{h\pi K}{K'}} e^{\pm \frac{\pi u}{2K'}} - e^{-\frac{h\pi K}{K'}} e^{\mp \frac{\pi u}{2K'}}}{e^{\frac{h\pi K}{K'}} e^{\pm \frac{\pi u}{2K'}} + e^{-\frac{h\pi K}{K'}} e^{\mp \frac{\pi u}{2K'}}}.$$

Put

(6)* $$q = e^{-\frac{\pi K'}{K}}, \quad q' = e^{-\frac{\pi K}{K'}},$$

and the last expression becomes

$$\operatorname{sn} \frac{\pi}{2K'}(2hK \pm u) = \frac{q'^{-h} e^{\pm \frac{\pi u}{2K'}} - q'^{h} e^{\mp \frac{\pi u}{2K'}}}{q'^{-h} e^{\pm \frac{\pi u}{2K'}} + q'^{h} e^{\mp \frac{\pi u}{2K'}}};$$

$$\operatorname{sn} \frac{\pi}{2K'}(2hK + u) \operatorname{sn} \frac{\pi}{2K'}(2hK - u)$$

$$= \frac{q'^{-h} e^{\frac{\pi u}{2K'}} - q'^{h} e^{-\frac{\pi u}{2K'}}}{q'^{-h} e^{\frac{\pi u}{2K'}} + q'^{h} e^{-\frac{\pi u}{2K'}}} \cdot \frac{q'^{-h} e^{-\frac{\pi u}{2K'}} - q'^{h} e^{\frac{\pi u}{2K'}}}{q'^{-h} e^{-\frac{\pi u}{2K'}} + q'^{h} e^{\frac{\pi u}{2K'}}}$$

$$= \frac{q'^{-2h} + q'^{2h} - \left(e^{\frac{\pi u}{K'}} + e^{-\frac{\pi u}{K'}}\right)}{q'^{-2h} + q'^{2h} + \left(e^{\frac{\pi u}{K'}} + e^{-\frac{\pi u}{K'}}\right)}.$$

From plane trigonometry we have the equations

$$\frac{e^x - e^{-x}}{e^x + e^{-x}} = -i \tan ix, \quad e^x + e^{-x} = 2 \cos ix;$$

where $i = \sqrt{-1}$: which gives

$$\operatorname{sn} \frac{\pi u}{2K'} = -i \tan \frac{\pi i u}{2K'}; \qquad \text{(Mod. 1;)}$$

$$\operatorname{sn} \frac{\pi}{2K'}(2hK + u) \operatorname{sn} \frac{\pi}{2K'}(2hK - u)$$

$$= \frac{q'^{-2h} + q'^{2h} - 2\cos\dfrac{\pi i u}{K'}}{q'^{-2h} + q'^{2h} + 2\cos\dfrac{\pi i u}{K'}}$$

$$= \frac{1 - 2q'^{2h}\cos\dfrac{\pi i u}{K'} + q'^{4h}}{1 + 2q'^{2h}\cos\dfrac{\pi i u}{K'} + q'^{4h}}.$$

From eq. (10), Chap. III, we have

$$\operatorname{sn}(u, k) = -i \operatorname{tn}(iu, k').$$

Substituting these values in eq. (6), we have

$$\operatorname{tn}(iu, k') = A' \tan \frac{\pi i u}{2K'} \left[\Pi\right] \frac{1 - 2q'^{2h}\cos\dfrac{\pi i u}{K'} + q'^{4h}}{1 + 2q'^{2h}\cos\dfrac{\pi i u}{K'} + q'^{4h}}.$$

Now in place of the series of moduli k', k_0' and the corresponding complete integral K', we are at liberty to substitute the parallel series of moduli k, k_0 and the corresponding complete integral K; calling the new integral u, we have

(8) $$\operatorname{tn}(u, k) = A \tan \frac{\pi u}{2K} \Pi \frac{1 - 2q^{2h}\cos\dfrac{\pi u}{K} + q^{4h}}{1 + 2q^{2h}\cos\dfrac{\pi u}{K} + q^{4h}}$$

$$= A \tan \frac{\pi u}{2K} \cdot \frac{1 - 2q^2 \cos \frac{\pi u}{K} + q^4}{1 + 2q^2 \cos \frac{\pi u}{K} + q^4}$$

$$\cdot \frac{1 - 2q^4 \cos \frac{\pi u}{K} + q^8}{1 + 2q^4 \cos \frac{\pi u}{K} + q^8}$$

$$\cdot \frac{1 - 2q^6 \cos \frac{\pi u}{K} + q^{12}}{1 + 2q^6 \cos \frac{\pi u}{K} + q^{12}} \cdots,$$

where

(9) $\qquad A = (1 + k_0)(1 + k_{02})^2(1 + k_{03})^4(1 + k_{04})^8 \cdots$

Now in equation (6) put $u + K$ for u, and we have, since [eq. (11), Chap. III] $\operatorname{sn}(u + K) = \frac{\operatorname{cn} u}{\operatorname{dn} u}$,

$$\frac{\operatorname{cn} u}{\operatorname{dn} u} = A' \operatorname{sn} \frac{\pi(u+K)}{2K'} \left[\Pi \right] \operatorname{sn} \frac{\pi}{2K'}[(2h+1)K + u]$$

$$\cdot \operatorname{sn} \frac{\pi}{2K'}[(2h-1)K - u]. \qquad \text{(Mod. 1.)}$$

Now from $2h - 1$ and $2h + 1$ we have the following series of numbers respectively:

$2h - 1$:   1, 3, 5, 7, 9, etc.
$2h + 1$:   3, 5, 7, 9, etc.

It will be observed that the factor outside of the sign $[\Pi]$, viz., $\sin \operatorname{am} \frac{\pi(u+K)}{2K'}$, would, if placed under the sign $[\Pi]$, supply the missing first term of the second series. Hence, placing this factor within the sign, we have

(10) $$\frac{\operatorname{cn} u}{\operatorname{dn} u} = A'[\Pi] \operatorname{sn} \frac{\pi}{2K'}[(2h-1)K+u]$$

$$\cdot \operatorname{sn} \frac{\pi}{2K'}[(2h-1)K-u]. \qquad \text{(Mod. 1.)}$$

Comparing this with equation (7), we see that the factors herein differ from those in equation (7) only in having $2h-1$ in place of $2h$; hence we have

$$\operatorname{sn} \frac{\pi}{2K'}[(2h-1)K+u] \operatorname{sn} \frac{\pi}{2K'}[(2h-1)K-u]$$

$$= \frac{1 - 2q'^{2h-1} \cos \frac{\pi i u}{K'} + q'^{4h-2}}{1 + 2q'^{2h-1} \cos \frac{\pi i u}{K'} + q'^{4h-2}}. \qquad \text{(Mod. 1.)}$$

From eqs. (10), Chap. III, we have

$$\frac{\operatorname{cn}(u, k)}{\operatorname{dn}(u, k)} = \frac{1}{\operatorname{dn}(iu, k')};$$

whence eq. (10) becomes

(11) $$\frac{1}{\operatorname{dn}(iu, k')} = A'[\Pi] \frac{1 - 2q'^{2h-1} \cos \frac{\pi i u}{K'} + q'^{4h-2}}{1 + 2q'^{2h-1} \cos \frac{\pi i u}{K'} + q'^{4h-2}};$$

and when in place of iu, k', K', q', A', we substitute u, k, K, q and A, and invert the equation, we have

(12) $$\operatorname{dn}(u, k) = \frac{1}{A}[\Pi] \frac{1 + 2q^{2h-1} \cos \frac{\pi u}{K} + q^{4h-2}}{1 - 2q^{2h-1} \cos \frac{\pi u}{K} + q^{4h-2}}.$$

Bearing in mind the remarkable property (Chap. III, p. 29) that the functions sn u and dn u approach infinity for the same value of u, we see that both these functions, except as to the factor independent of u, must have the same denominator. Furthermore, since sn u and tn u disappear for the same value of u, they must, except for the independent factor, have the same numerator. Hence, indicating by B a new quantity, dependent upon k but independent of u, we have

$$(13) \quad \operatorname{sn}(u, k) = B \sin \frac{\pi u}{2K} \left[\Pi \right] \frac{1 - 2q^{2h} \cos \frac{\pi u}{K} + q^{4h}}{1 - 2q^{2h-1} \cos \frac{\pi u}{K} + q^{4h-2}};$$

and since

$$\operatorname{cn} u = \frac{\operatorname{sn} u}{\operatorname{tn} u},$$

we also have, from eqs. (8) and (13),

$$(14) \quad \operatorname{cn}(u, k) = \frac{B}{A} \cos \frac{\pi u}{2K} \left[\Pi \right] \frac{1 + 2q^{2h} \cos \frac{\pi u}{K} + q^{4h}}{1 - 2q^{2h-1} \cos \frac{\pi u}{K} + q^{4h-2}}.$$

Collecting these results, we have the following equations:

$$(15) \quad \operatorname{sn}(u, k) = B \sin \frac{\pi u}{2K} \left[\Pi \right] \frac{1 - 2q^{2h} \cos \frac{\pi u}{K} + q^{4h}}{1 - 2q^{2h-1} \cos \frac{\pi u}{K} + q^{4h-2}},$$

$$(16) \quad \operatorname{cn}(u, k) = \frac{B}{A} \cos \frac{\pi u}{2K} \left[\Pi \right] \frac{1 + 2q^{2h} \cos \frac{\pi u}{K} + q^{4h}}{1 - 2q^{2h-1} \cos \frac{\pi u}{K} + q^{4h-2}},$$

$$(17) \quad \mathrm{dn}\,(u, k) = \frac{1}{A}\Big[\Pi\Big]\frac{1 + 2q^{2h-1}\cos\frac{\pi u}{K} + q^{4h-2}}{1 - 2q^{2h-1}\cos\frac{\pi u}{K} + q^{4h-2}}.$$

To ascertain the values of A and B, we proceed as follows:

In eq. (17) we make $u = 0$, whence, by eq. (13), Chap. II, we have

$$1 = \frac{1}{A}\Big[\Pi\Big]\Big(\frac{1 + q^{2h-1}}{1 - q^{2h-1}}\Big)^2;$$

whence

$$(18) \quad \frac{1}{A} = \Big[\Pi\Big]\Big(\frac{1 - q^{2h-1}}{1 + q^{2h-1}}\Big)^2.$$

In equation (17), making $u = K$, we get, by equation (1), Chap. III,

$$k' = \frac{1}{A}\Big[\Pi\Big]\Big(\frac{1 - q^{2h-1}}{1 + q^{2h-1}}\Big)^2 = \frac{1}{A^2};$$

$$(19) \quad \therefore \frac{1}{A} = \sqrt{k'}.$$

We have identically

$$1 = B\frac{1}{B} = B\frac{\frac{1}{A}}{\frac{B}{A}} = B\frac{\sqrt{k'}}{\frac{B}{A}};$$

whence

$$\frac{B}{A} = B\sqrt{k'}.$$

To calculate B, put $e^{\frac{i\pi u}{2K}} = v$; if we change $\frac{\pi u}{2K}$ into

$\frac{\pi u}{2K} + \frac{i\pi K'}{2K}$, v will change into $v\sqrt{q}$, and sn u will become, by eq. (14), Chap. III,

$$\text{sn}(u + iK') = \frac{1}{k\,\text{sn}\,u}.$$

Now, replacing $\sin\frac{\pi u}{2K}$ and $\cos\frac{\pi u}{K}$ by their exponential values, and observing that

$$1 - 2q^n \cos\frac{\pi u}{K} + q^{2n} = (1 - q^n v^2)(1 - q^n v^{-2}),$$

we have

$$\text{sn}\,u = \frac{B}{2} \cdot \frac{v - v^{-1}}{\sqrt{-1}} \cdot \frac{[\Pi](1 - q^{2h}v^2)(1 - q^{2h}v^{-2})}{[\Pi](1 - q^{2h-1}v^2)(1 - q^{2h-1}v^{-2})}.$$

Changing u into $u + iK'$, and consequently v into $v\sqrt{q}$, we have

$$\frac{1}{k\,\text{sn}\,u} = \frac{B}{2} \cdot \frac{v\sqrt{q} - v^{-1}\sqrt{q^{-1}}}{\sqrt{-1}} \cdot \frac{[\Pi](1 - q^{2h+1}v^2)(1 - q^{2h-1}v^{-2})}{[\Pi](1 - q^{2h}v^2)(1 - q^{2h-2}v^{-2})}.$$

Multiplying these equations together, member by member, and observing that

$$v\sqrt{q} - v^{-1}\sqrt{q^{-1}} = \frac{1 - qv^2}{-v\sqrt{q}},$$

$$v - v^{-1} = v(1 - v^{-2}),$$

we get

$$\frac{1}{k} = \frac{B^2}{4} \cdot \frac{1 - qv^2}{v\sqrt{q}} \cdot v(1 - v^{-2}) \cdot \frac{[\Pi](1 - q^{2h+1}v^2)(1 - q^{2h}v^{-2})}{[\Pi](1 - q^{2h-1}v^2)(1 - q^{2h-2}v^{-2})}$$

DEVELOPMENT INTO FACTORS.

$$= \frac{B^2}{4\sqrt{q}}(1-qv^2)v(1-v^{-2})\frac{(1-q^3v^2)(1-q^5v^2)\cdots}{(1-qv^2)(1-q^3v^2)\cdots}$$

$$\cdot \frac{(1-q^3v^{-2})(1-q^5v^{-2})\cdots}{(1-v^{-2})(1-q^2v^{-2})\cdots}$$

$$= \frac{B^2}{4}\cdot\frac{1}{\sqrt{q}}.$$

$$\therefore B = \frac{2\sqrt[4]{q}}{\sqrt{k}};$$

whence

$$\frac{B}{A} = 2\sqrt[4]{q}\sqrt{\frac{k'}{k}}.$$

Substituting these values in eqs. (15), (16), and (17), we have

$$(20)\quad \operatorname{sn}(u, k) = \frac{2\sqrt[4]{q}}{\sqrt{k}}\sin\frac{\pi u}{2K}\left[\Pi\right]\frac{1-2q^{2h}\cos\frac{\pi u}{K}+q^{4h}}{1-2q^{2h-1}\cos\frac{\pi u}{K}+q^{4h-2}};$$

$$(21)\quad \operatorname{cn}(u, k) = \frac{2\sqrt{k'}\sqrt[4]{q}}{\sqrt{k}}\cos\frac{\pi u}{2K}\left[\Pi\right]\frac{1+2q^{2h}\cos\frac{\pi u}{K}+q^{4h}}{1-2q^{2h-1}\cos\frac{\pi u}{K}+q^{4h-2}};$$

$$(22)\quad \operatorname{dn}(u, k) = \sqrt{k'}\left[\Pi\right]\frac{1+2q^{2h-1}\cos\frac{\pi u}{K}+q^{4h-2}}{1-2q^{2h-1}\cos\frac{\pi u}{K}+q^{4h-2}}.$$

CHAPTER VIII.

THE Θ FUNCTION.

We will indicate the denominator in eq. (20), Chap. VII, by $\phi(u)$, thus:

(1) $$\phi(u) = [\Pi](1 - 2q^{2h-1}\cos\frac{\pi u}{K} + q^{4h-2}).$$

We will now develop this into a series consisting of the cosines of the multiples of $\frac{\pi u}{K}$. Put $\frac{\pi u}{2K} = x$, whence

$$2\cos\frac{\pi u}{K} = (e^{2ix} + e^{-2ix});$$

but

$$1 - 2q^{2h-1}\cos\frac{\pi u}{K} + q^{4h-2} = (1 - q^{2h-1}e^{2ix})(1 - q^{2h-1}e^{-2ix}),$$

and therefore

(2) $$\phi(u) = (1 - qe^{2ix})(1 - q^3 e^{2ix})(1 - q^5 e^{2ix})\ldots$$
$$(1 - qe^{-2ix})(1 - q^3 e^{-2ix})(1 - q^5 e^{-2ix})\ldots$$

Putting now $u + 2iK'$ instead of u, we have

$$x_1 = \frac{\pi(u + 2iK')}{2K} = x + \frac{\pi iK'}{K},$$

$$2ix_1 = 2ix - \frac{2\pi K'}{K};$$

and
$$e^{2ix_1} = q^2 e^{2ix},$$
$$e^{-2ix_1} = \frac{1}{q^2} e^{-2ix}.$$

From these we have
$$\phi(u+2iK') = -\frac{1}{q}e^{-2ix}(1-qe^{2ix})(1-q^3e^{2ix})\ldots$$
$$(1-qe^{-2ix})(1-q^3e^{-2ix})\ldots;$$

whence
$$\phi(u+2iK') = -\frac{1}{q}e^{-2ix}\phi(u),$$

or

(3) $$\phi(u+2iK') = -q^{-1}e^{-\frac{\pi i u}{K}}\phi(u).$$

Now put

(4) $$\phi(u) = A + B\cos\frac{\pi u}{K} + C\cos\frac{2\pi u}{K} + D\cos\frac{3\pi u}{K} + \text{etc.}$$

Since
$$\cos\frac{\pi u}{K} = \tfrac{1}{2}(e^{2ix} + e^{-2ix}),$$

this becomes

(5) $$\phi(u) = A + \tfrac{1}{2}Be^{2ix} + \tfrac{1}{2}Ce^{4ix} + \tfrac{1}{2}De^{6ix} + \ldots$$
$$+ \tfrac{1}{2}Be^{-2ix} + \tfrac{1}{2}Ce^{-4ix} + \tfrac{1}{2}De^{-6ix} + \ldots;$$

whence

(6) $$-\frac{1}{q}e^{-2ix}\phi(u) = -\frac{A}{q}e^{-2ix} - \frac{B}{2q} - \frac{C}{2q}e^{2ix} - \frac{D}{2q}e^{4ix} - \ldots$$
$$- \frac{B}{2q}e^{-4ix} - \frac{C}{2q}e^{-6ix} - \frac{D}{2q}e^{-8ix} - \ldots$$

Now in equation (5) put $u + 2iK'$ in place of u, remembering that e^{2ix} and e^{-2ix} are thereby changed respectively into $q^2 e^{2ix}$ and $q^{-2} e^{-2ix}$, and we have

(7) $\quad \phi(u + 2iK') = A + \dfrac{Bq^2}{2} e^{2ix} + \dfrac{Cq^4}{2} e^{4ix} + \dfrac{Dq^6}{2} e^{6ix} + \ldots$

$$+ \dfrac{B}{2q^2} e^{-2ix} + \dfrac{C}{2q^4} e^{-4ix} + \ldots$$

Since equations (6) and (7) are equal, we have

$$-\dfrac{B}{2q} = A, \qquad\qquad B = -2qA;$$

$$-\dfrac{C}{2q} = \dfrac{Bq^2}{2}, \qquad\qquad C = +2q^4 A;$$

$$-\dfrac{D}{2q} = \dfrac{Cq^4}{2}, \qquad\qquad D = -2q^9 A;$$

$$\cdots\cdots\cdots\qquad\qquad\cdots\cdots\cdots$$

whence

(8) $\quad \begin{cases} [\Pi](1 - 2q^{2k-1} \cos \dfrac{\pi u}{K} + q^{4k-2}) \\[4pt] = A(1 - 2q \cos \dfrac{\pi u}{K} + 2q^4 \cos \dfrac{2\pi u}{K} - 2q^9 \cos \dfrac{3\pi u}{K} \\[4pt] \qquad\qquad + 2q^{16} \cos \dfrac{4\pi u}{K} - \ldots). \end{cases}$

The series in the second member has been designated by Jacobi and subsequent writers by $\Theta(u)$, thus:

(9) $\quad \Theta(u) = 1 - 2q \cos \dfrac{\pi u}{K} + 2q^4 \cos \dfrac{2\pi u}{K} - \ldots$

CHAPTER IX.

THE Θ AND H FUNCTIONS.

In equation (20), Chap. VII, viz.,

$$\operatorname{sn}(u, k) = \frac{2\sqrt[4]{q}}{\sqrt{k}} \sin \frac{\pi u}{2K} \left[\Pi\right] \frac{1 - 2q^{2h}\cos\frac{u\pi}{K} + q^{4h}}{1 - 2q^{2h-1}\cos\frac{\pi u}{K} + q^{4h-2}},$$

the numerator and the denominator have been considered separately by Jacobi, who gave them a special notation and developed from them a theory second only in importance to the elliptic functions themselves.

Put [see equation (8), Chap. VIII]

(1) $$\Theta(u) = \frac{1}{A} \left[\Pi\right] (1 - 2q^{2h-1} \cos \frac{\pi u}{K} + q^{4h-2}).$$

(2) $$H(u) = 2 \frac{1}{A} \sqrt[4]{q} \sin \frac{\pi u}{2K} \left[\Pi\right] (1 - 2q^{2h} \cos \frac{\pi u}{K} + q^{4h});$$

A being a constant whose value is to be determined later. From these we have

(3) $$\operatorname{sn}(u, k) = \frac{1}{\sqrt{k}} \cdot \frac{H(u)}{\Theta(u)}.$$

The functions sn u and cn u can also be expressed in terms of the new functions; thus we have

$$(4) \quad \text{cn}(u, k) = \sqrt{\frac{k'}{k}} \cdot 2\sqrt[4]{q} \cos\frac{\pi u}{2K} \left[\Pi\right] \frac{1 + 2q^{2h}\cos\frac{\pi u}{K} + q^{4h}}{1 - 2q^{2h-1}\cos\frac{\pi u}{K} + q^{4h-2}};$$

or, since $\sin x = \cos\left(x + \frac{\pi}{2}\right)$ and $\cos x = -\cos\left(x + \frac{\pi}{2}\right)$,

and putting $u = \frac{2Kx}{\pi}$,

$$\text{cn}\left(\frac{2Kx}{\pi}, k\right) = \sqrt{\frac{k'}{k}} \frac{H\left[\frac{2K}{\pi}\left(x + \frac{\pi}{2}\right)\right]}{\Theta\left(\frac{2Kx}{\pi}\right)}$$

$$= \sqrt{\frac{k'}{k}} \frac{H\left[\frac{2Kx}{\pi} + K\right]}{\Theta\left(\frac{2Kx}{\pi}\right)}.$$

Replacing $\frac{2Kx}{\pi}$ by its value, u, we have

$$(5) \quad \text{cn}(u, k) = \sqrt{\frac{k'}{k}} \frac{H(u + K)}{\Theta(u)}.$$

Furthermore,

$$(6) \quad \text{dn}(u, k) = \sqrt{k'}\left[\Pi\right]\frac{1 + 2q^{2h-1}\cos\frac{\pi u}{K} + q^{4h-2}}{1 - 2q^{2h-1}\cos\frac{\pi u}{K} + q^{4h-2}}$$

gives in the same manner

$$\operatorname{dn}\frac{2Kx}{\pi} = \sqrt{k'}\,\frac{\Theta\left[\frac{2K}{\pi}\left(x+\frac{\pi}{2}\right)\right]}{\Theta\left(\frac{2Kx}{\pi}\right)},$$

or

(7) $$\operatorname{dn}(u, k) = \sqrt{k'}\,\frac{\Theta(u+K)}{\Theta(u)}.$$

If we put

(8) $$H(u+K) = H_1(u),$$

(9) $$\Theta(u+K) = \Theta_1(u),$$

the three elliptic functions can be expressed by the following formulas:

(10) $$\operatorname{sn}(u, k) = \frac{1}{\sqrt{k}} \cdot \frac{H(u)}{\Theta(u)};$$

(11) $$\operatorname{cn}(u, k) = \sqrt{\frac{k'}{k}} \cdot \frac{H_1(u)}{\Theta(u)};$$

(12) $$\operatorname{dn}(u, k) = \sqrt{k'}\,\frac{\Theta_1(u)}{\Theta(u)}.$$

These functions Θ and H can be expressed in terms of each other. By definition,

$$H(u) = 2C\sqrt[4]{q}\sin\frac{\pi u}{2K}\left[\Pi\right]\left(1 - 2q^{2h}\cos\frac{\pi u}{K} + q^{4h}\right);$$

but

$$1 - 2q^h \cos \frac{\pi u}{K} + q^{2h} = \left(1 - q^h e^{\frac{\pi u \sqrt{-1}}{K}}\right)\left(1 - q^h e^{-\frac{\pi u \sqrt{-1}}{K}}\right),$$

$$\sin \frac{\pi u}{2K} = \frac{e^{\frac{\pi i u}{2K}} - e^{-\frac{\pi i u}{2K}}}{2\sqrt{-1}}$$

$$= e^{-\frac{\pi i u}{2K}} \frac{1 - e^{\frac{\pi i u}{K}}}{2} \sqrt{-1},$$

and consequently

(13) $\quad H(u) = C \sqrt[4]{q} e^{-\frac{\pi i u}{2K}} \sqrt{-1}\left(1 - e^{\frac{\pi i u}{K}}\right)\left(1 - q^2 e^{-\frac{\pi i u}{K}}\right)\left(1 - q^2 e^{\frac{\pi i u}{K}}\right)\ldots$

Now, changing u into $u + iK'$, and remembering that $e^{\frac{-\pi K'}{K}} = q$, we have

(14) $\quad H(u + iK')$

$$= Cq^{-\frac{1}{4}} e^{\frac{-\pi i u}{2K}} \sqrt{-1}\left(1 - qe^{\frac{\pi i u}{K}}\right)\left(1 - qe^{-\frac{\pi i u}{K}}\right)\left(1 - q^3 e^{\frac{\pi i u}{K}}\right)\left(1 - q^3 e^{-\frac{\pi i u}{K}}\right)\ldots;$$

and reuniting the factors two by two, this becomes

(15) $\quad H(u + iK')$

$$= C\sqrt{-1}\, q^{-\frac{1}{4}} e^{-\frac{\pi i u}{2K}} \left(1 - 2q \cos \frac{\pi u}{K} + q^2\right)\left(1 - 2q^3 \cos \frac{\pi u}{K} + q^6\right)\ldots;$$

and finally, according to equation (1),

(16) $\qquad H(u + iK') = \sqrt{-1}\, q^{-\frac{1}{4}} e^{-\frac{\pi i u}{2K}} \Theta(u).$

In the same manner, we can get

(17) $$\Theta(u + iK') = \sqrt{-1}q^{-\frac{1}{4}}e^{-\frac{\pi i u}{2K}}H(u).$$

Substituting $u + 2K$ for u in equations (1) and (2), we get

(18) $$\Theta(u + 2K) = \Theta(u),$$

(19) $$H(u + 2K) = -H(u),$$

since $\cos\frac{\pi}{K}(u + 2K) = \cos\frac{\pi u}{K}$ and $\sin\frac{\pi}{2K}(u+2K) = -\sin\frac{\pi u}{2K}$.

The comparison of these four equations with equations (10), (11), and (12) shows the periodicity of the elliptic functions. For example, comparing eqs. (10) and (16) and (17), we see that changing u into $u + iK'$ simply multiplies the numerator and denominator of the second member of eq. (10) by the same number, and does not change their ratio.

The addition of $2K$ changes the sign of the function, but not its value.

We will define Θ_1 and H_1 as follows:

(20) $$\Theta_1(x) = \Theta(x + K);$$

(21) $$H_1(x) = H(x + K).$$

Hence we get, from equation (17),

$$\Theta_1(x + iK') = \Theta(x + iK' + K) = \Theta(x + K + iK')$$

$$= iH(x + K)e^{-\frac{i\pi}{4K}(2x + 2K + iK')}$$

$$= iH_1(x)e^{-\frac{i\pi}{4K}(2x + iK')}(-\sqrt{-1}),$$

since $e^{-\frac{i\pi}{2}} = \cos\frac{\pi}{2} - \sqrt{-1}\sin\frac{\pi}{2} = -\sqrt{-1};$

74   ELLIPTIC FUNCTIONS.

whence

(22) $$\Theta_1(x+iK') = H_1(x)e^{-\frac{i\pi}{4K}(2x+iK')}.$$

In a similar manner we get

(22)* $$H_1(x+iK') = \Theta_1(x)e^{-\frac{i\pi}{4K}(2x+iK')}.$$

In eq. (9), Chap. VIII, put $u = \frac{2Kz}{\pi}$, and we get

(23) $$\Theta\left(\frac{2Kz}{\pi}\right) = 1 - 2q\cos 2z + 2q^4\cos 4z - \ldots$$

Now, in this equation, changing z into $z + \frac{\pi}{2}$, and observing eq. (20), we get

(24) $$\Theta_1\left(\frac{2Kz}{\pi}\right) = 1 + 2q\cos 2z + 2q^4\cos 4z + \ldots$$

Applying eq. (22) to this, we have

$$H_1\left(\frac{2Kz}{\pi}\right) = \Theta_1\left(\frac{2K}{\pi}\left(z + \frac{\pi i K'}{2K}\right)\right) e^{\frac{\pi i}{4K}\left(\frac{4Kz}{\pi} + iK'\right)}$$

$$= \Theta_1\left(\frac{2K}{\pi}\left(z + \frac{\pi i K'}{2K}\right)\right) e^{iz} q^{\frac{1}{4}}$$

$$= e^{iz} q^{\frac{1}{4}}\left[1 + 2q\cos 2\left(z + \frac{\pi i K'}{2K}\right) + 2q^4\cos 4\left(z + \frac{\pi i K'}{2K}\right) + \ldots\right]$$

$$= e^{iz} q^{\frac{1}{4}}\left[1 + q\left(e^{2i\left(z + \frac{\pi i K'}{2K}\right)} + e^{-2i\left(z + \frac{\pi i K'}{2K}\right)}\right)\right.$$

$$\left. + q^4\left(e^{4i\left(z + \frac{\pi i K'}{2K}\right)} + e^{-4i\left(z + \frac{\pi i K'}{2K}\right)}\right) + \ldots\right]$$

$$= e^{iz}q^{\frac{1}{4}}[1 + q(qe^{2iz} + q^{-1}e^{-2iz}) + q^4(q^2e^{4iz} + q^{-2}e^{-4iz}) + \ldots]$$
$$= e^{iz}q^{\frac{1}{4}}[1 + q^2e^{2iz} + q^6e^{4iz} + \ldots$$
$$\qquad\qquad\qquad + e^{-2iz} + q^6e^{-4iz} + \ldots]$$
$$= q^{\frac{1}{4}}[e^{iz} + q^2e^{3iz} + q^6e^{5iz} + \ldots$$
$$\qquad\qquad\qquad + e^{-iz} + q^2e^{-3iz} + q^6e^{-5iz} + \ldots]$$
$$= 2q^{\frac{1}{4}}[\cos z + q^2 \cos 3z + q^6 \cos 5z + \ldots];$$

whence

$$(25) \quad H_1\left(\frac{2Kz}{\pi}\right) = 2\sqrt[4]{q}\cos z + 2\sqrt[4]{q^9}\cos 3z + 2\sqrt[4]{q^{25}}\cos 5z + \ldots$$

In this equation, changing z into $z - \dfrac{\pi}{2}$, and applying eq. (21), we get

$$(26) \quad H\left(\frac{2Kz}{\pi}\right) = 2\sqrt[4]{q}\sin z - 2\sqrt[4]{q^9}\sin 3z + 2\sqrt[4]{q^{25}}\sin 5z - \ldots,$$

since

$$H_1\left(\frac{2Kz}{\pi}\right) = H\left(\frac{2Kz}{\pi} + K\right).$$

We will now determine the constant A of eq. (8), Chap. VIII, and eqs. (1) and (2) of this chapter. Denote A by $f(q)$, and we have

$$(26)^* \quad [\Pi](1 - 2q^{2h-1}\cos\frac{\pi u}{K} + q^{4h-2}) = f(q)\Theta(u).$$

Substituting herein $u = 0$ and $u = \dfrac{K}{2}$, we have

$$[\Pi](1 - q^{2h-1})^2 = f(q)\Theta(0);$$
$$[\Pi](1 + q^{4h-2}) = f(q)\Theta\left(\frac{K}{2}\right).$$

From eq. (9), Chap. VIII, we get

(27) $$\Theta(0) = 1 - 2q + 2q^4 - 2q^9 + 2q^{16} - \ldots;$$

(28) $$\Theta\left(\frac{K}{2}\right) = 1 - 2q^4 + 2q^{16} - 2q^{36} + 2q^{64} - \ldots;$$

from which we see that $\Theta(0)$ is changed into $\Theta\left(\frac{K}{2}\right)$ when we put q^4 in place of q.

Whence
$$[\Pi](1 - q^{8h-4})^2 = f(q^4)\Theta\left(\frac{K}{2}\right);$$

and therefore
$$\frac{f(q)}{f(q^4)} = [\Pi]\frac{1 + q^{4h-2}}{(1 - q^{8h-4})^2}$$

(29) $$= [\Pi]\frac{1}{(1 - q^{8h-4})(1 - q^{4h-2})}.$$

Now, the expressions $4h - 2$, $8h - 4$, and $8h$ give the following series of numbers:

$4h - 2$,   2, 6, 10, 14, 18, 22, 26, 30, 34;
$8h - 4$,   4,   12,   20,   28,   36;
$8h$,   8,   16,   24,   32.

Hence, the three expressions taken together contain all the even numbers, and

$$[\Pi](1 - q^{8h-4})(1 - q^{4h-2})(1 - q^{8h}) = [\Pi](1 - q^{2h}).$$

Therefore, multiplying eq. (29) by

$$[\Pi]\frac{1 - q^{8h}}{1 - q^{8h}},$$

we have

$$\frac{f(q)}{f(q^4)} = [\Pi]\frac{1 - q^{8h}}{1 - q^{2h}}.$$

Now in this equation, by successive substitutions of q^4 for q, we get

$$\frac{f(q^4)}{f(q^{16})} = \left[\Pi\right]\frac{1-q^{32k}}{1-q^{8k}};$$

$$\frac{f(q^{16})}{f(q^{64})} = \left[\Pi\right]\frac{1-q^{128k}}{1-q^{32k}};$$

$$\frac{f(q^{64})}{f(q^{256})} = \left[\Pi\right]\frac{1-q^{512k}}{1-q^{128k}};$$

. . . .

Now q being less than 1, q^n tends towards the limit 0 as n increases, and consequently $1 - q^n$ tends towards the limit 1. Also, from eq. (8), Chap. VIII, we see that $f(0) = 1$. Hence, multiplying the above equations together member by member, we have

(30) $$f(q) = \left[\Pi\right]\frac{1}{1-q^{2k}},$$

or

(31) $$A = \frac{1}{(1-q^2)(1-q^4)(1-q^6)\ldots}.$$

Substituting this value in equation (8), Chap. VIII, we have, after making $u = 0$,

$$(1-q)^2(1-q^3)^2(1-q^5)^2 \ldots = \frac{1 - 2q + 2q^4 - 2q^9 + \ldots}{(1-q^2)(1-q^4)(1-q^6)\ldots}$$

$$= \frac{\Theta(0)}{(1-q^2)(1-q^4)(1-q^6)\ldots}.$$

(See equation (9), Chap. VIII.)

Transposing one of the series of products from the left-hand member, we get

$$(1-q)(1-q^3)\ldots = \frac{\Theta(0)}{(1-q)(1-q^2)(1-q^3)(1-q^4)\ldots}.$$

Introducing on both sides of the equation the factors $1-q^2$, $1-q^4$, $1-q^6$, etc., we get

$$(1-q)(1-q^2)(1-q^3)(1-q^4)\cdots$$
$$=\Theta(0)\frac{1-q^2}{1-q}\cdot\frac{1-q^4}{1-q^2}\cdot\frac{1-q^6}{1-q^3}\cdot\frac{1-q^8}{1-q^4}\cdots$$
$$=\Theta(0)(1+q)(1+q^2)(1+q^3)\cdots;$$

whence

(32) $$\Theta(0)=\frac{(1-q)(1-q^2)(1-q^3)}{(1+q)(1+q^2)(1+q^3)}.$$

Resuming equation (20), Chap. VII, and dividing both members of the equation by u, we have

$$\frac{\operatorname{sn} u}{u}=\frac{2\sqrt[4]{q}}{\sqrt{k}}\frac{\sin\frac{\pi u}{2K}}{u}\left[\Pi\right]\frac{1-2q^{2h}\cos\frac{\pi u}{K}+q^{4h}}{1-2q^{2h-1}\cos\frac{\pi u}{K}+q^{4h-2}}.$$

This, for $u=0$, since the limiting value of $\dfrac{\operatorname{sn} u}{u}$ for $u=0$ is 1, and of $\dfrac{\sin\frac{\pi u}{2K}}{u}$ for $x=0$ is $\dfrac{\pi}{2K}$, becomes

$$1=\frac{\sqrt[4]{q}}{\sqrt{k}}\cdot\frac{\pi}{K}\cdot\frac{(1-q^2)^2(1-q^4)^2(1-q^6)^2\cdots}{(1-q)^2(1-q^3)^2(1-q^5)^2\cdots};$$

or

(33) $$\frac{\sqrt{k}K}{\pi\sqrt[4]{q}}=\left[\frac{(1-q^2)(1-q^4)(1-q^6)\cdots}{(1-q)(1-q^3)(1-q^5)\cdots}\right]^2.$$

Further, from equation (21), Chap. VII, for $u=0$, we have

(34) $$\frac{\sqrt{k}}{2\sqrt{k'}\sqrt[4]{q}}=\left[\frac{(1+q^2)(1+q^4)(1+q^6)\cdots}{(1-q)(1-q^3)(1-q^5)\cdots}\right]^2.$$

The quotient of these two equations gives

$$(35) \quad \frac{2\sqrt{k'}K}{\pi} = \left[\frac{(1-q^2)(1-q^4)(1-q^6)\cdots}{(1+q^2)(1+q^4)(1+q^6)\cdots}\right]^2;$$

or, substituting the value of $\sqrt{k'}$ from eqs. (18) and (19), Chap. VII,

$$(36) \quad \frac{2k'K}{\pi} = \left[\frac{(1-q)(1-q^2)(1-q^3)\cdots}{(1+q)(1+q^2)(1+q^3)\cdots}\right]^2.$$

Comparing this with equation (32), we easily get

$$(37) \quad \Theta(0) = \sqrt{\frac{2k'K}{\pi}}.$$

From equation (9), Chap. VIII, making $u = K$, we get

$$(38) \quad \Theta(K) = 1 + 2q + 2q^4 + 2q^9 + 2q^{16} + \cdots$$

Making $z = 0$ in equation (24), Chap. IX, we have

$$(39) \quad \Theta_1(0) = 1 + 2q + 2q^4 + 2q^9 + \cdots$$

This might also have been derived from eq. (38) by observing that

$$\Theta_1(0 + K) = \Theta_1(0) = \Theta(K).$$

Knowing $\Theta(0)$, it is easy to deduce $\Theta(K)$ and $H(K)$.

From equation (7) we have

$$\mathrm{dn}\, u = \sqrt{k'}\, \frac{\Theta(u+K)}{\Theta(u)}.$$

Making $u = 0$, we have, since $\mathrm{dn}(0) = 1$,

$$(40) \quad \Theta(K) = \frac{\Theta(0)}{\sqrt{k'}}.$$

From equation (5) we get, in the same manner,

(41) $$H(K) = \sqrt{\frac{k'}{k}}\, \Theta(0).$$

From eq. (12), Chap. IX, we have

(41)* $$\operatorname{dn} u = \sqrt{1 - k^2 \sin^2 \phi} = \sqrt{k'}\, \frac{\Theta_1(u)}{\Theta(u)};$$

and putting $x = \dfrac{\pi u}{2K}$, we have

(42) $$\frac{\operatorname{dn} u}{\sqrt{k'}} = \frac{1 + 2q \cos 2x + 2q^4 \cos 4x + 2q^9 \cos 6x + \ldots}{1 - 2q \cos 2x + 2q^4 \cos 4x - 2q^9 \cos 6x + \ldots}.$$

Putting

(42)* $$\frac{\operatorname{dn} u}{\sqrt{k'}} = \cot \gamma,$$

we have

$$\frac{\cot \gamma - 1}{\cot \gamma + 1} = \tan(45° - \gamma) = 2q\, \frac{\cos 2x + q^2(4\cos^3 2x - 3\cos 2x) + \ldots}{1 + q^4(4\cos^2 2x - 2)};$$

whence

(43) $$\cos 2x = \frac{\tan(45° - \gamma)[1 + q^4(4\cos^2 2x - 2)]}{2q} - q^2(4\cos^3 2x - 3\cos 2x) - \ldots,$$

and approximately,

(44) $$\cos 2x = \frac{\tan(45° - \gamma)}{2q}.$$

From equations (37) and (40), Chap. IX, we have

(45) $$x = \frac{u}{\Theta^2(K)};$$

whence

(46) $$u = x\Theta^2(K).$$

CHAPTER X.

ELLIPTIC INTEGRALS OF THE SECOND ORDER.

From Chap. I, equation (19), we have

$$E(k, \phi) = \int_0^\phi \sqrt{1 - k^2 \sin^2 \phi} \cdot d\phi = \int_0^\phi \Delta\phi \cdot d\phi.$$

From this we have

$$E(\phi) + E(\psi) = \int_0^\phi \Delta\phi \cdot d\phi + \int_0^\psi \Delta\phi \cdot d\phi.$$

Put

(1) $$E\phi + E\psi = S.$$

Differentiating, we get

(2) $$\Delta\phi \cdot d\phi + \Delta\psi \cdot d\psi = dS.$$

But we have, Chap. II, equation (2),

$$\frac{d\phi}{\Delta\phi} + \frac{d\psi}{\Delta\psi} = 0,$$

or

(3) $$\Delta\psi \cdot d\phi + \Delta\phi \cdot d\psi = 0.$$

Adding equations (2) and (3), we get

(4) $$(\Delta\phi + \Delta\psi)(d\phi + d\psi) = dS.$$

ELLIPTIC FUNCTIONS.

Substituting cos μ from eq. (5), in eq. (5)*, Chap. II, we get

(5) $$\begin{cases} \varDelta\phi = \dfrac{\sin\phi\cos\psi\varDelta\mu + \cos\phi\sin\psi}{\sin\mu}, \\ \varDelta\psi = \dfrac{\sin\psi\cos\phi\varDelta\mu + \cos\psi\sin\phi}{\sin\mu}; \end{cases}$$

whence

(6) $$\varDelta\phi \pm \varDelta\psi = \frac{\varDelta\mu \pm 1}{\sin\mu}\sin(\phi \pm \psi).$$

Substituting in equation (4), we have

(7) $$\begin{aligned} dS &= \frac{\varDelta\mu + 1}{\sin\mu}\sin(\phi + \psi)d(\phi + \psi) \\ &= -\frac{\varDelta\mu + 1}{\sin\mu}d\cos(\phi + \psi). \end{aligned}$$

Integrating equation (7), we have

$$E\phi + E\psi = \frac{\varDelta\mu + 1}{\sin\mu}[C - \cos(\phi + \psi)].$$

The constant of integration, C, is determined by making $\phi = 0$; in this case $\psi = \mu$, $E\phi = 0$, $E\psi = E\mu$, and $S = E\mu$; whence

$$E\mu = \frac{\varDelta\mu + 1}{\sin\mu}(C - \cos\mu),$$

and by subtraction,

$$E\phi + E\psi - E\mu = \frac{\varDelta\mu + 1}{\sin\mu}(\cos\mu - \cos\phi\cos\psi + \sin\phi\sin\psi).$$

But, Chap. II, eq. (5),

$$\cos\mu - \cos\phi\cos\psi = -\sin\phi\sin\psi\varDelta\mu;$$

whence

$$E\phi + E\psi - E\mu = \frac{1 - \Delta^2\mu}{\sin \mu} \sin \phi \sin \psi$$

whence

(8) $$E\phi + E\psi = E\mu + k^2 \sin \phi \sin \psi \sin \mu.$$

When $\phi = \psi$, we have

(9) $$E\mu = 2E\phi - k^2 \sin^2 \phi \sin \mu.$$

But in that case

(10) $$\cos \mu = \cos^2 \phi - \sin^2 \phi \Delta\mu;$$

whence

(11) $$\sin \phi = \sqrt{\frac{1 - \cos \mu}{1 + \Delta\mu}}.$$

Let ϕ, ϕ_1, ϕ_2, etc., be such values as will satisfy the equations

(12) $$E\phi = 2E\phi_1 - k^2 \sin^2 \phi_1 \sin \phi,$$
$$E\phi_1 = 2E\phi_2 - k^2 \sin^2 \phi_2 \sin \phi_1,$$
$$\text{etc.} \qquad \text{etc.}$$

Assume an auxiliary angle γ, such that

(13) $$\sin \gamma = k \sin \phi;$$

whence

$$\Delta\phi = \cos \gamma,$$

and Chap. IV, eq. (24),

(14) $$\sin \phi_1 = \frac{\sin \tfrac{1}{2}\phi}{\cos \tfrac{1}{2}\gamma}.$$

84   ELLIPTIC FUNCTIONS.

Applying eqs. (13) and (14) successively, we get

$$(15)\begin{cases} \sin\phi_{\frac{1}{2}} = \dfrac{\sin\frac{1}{2}\phi}{\cos\frac{1}{2}\gamma}, \quad \sin\gamma_{\frac{1}{2}} = k\sin\phi_{\frac{1}{2}}; \\[1em] \sin\phi_{\frac{1}{4}} = \dfrac{\sin\frac{1}{2}\phi_{\frac{1}{2}}}{\cos\frac{1}{2}\gamma_{\frac{1}{2}}}, \quad \sin\gamma_{\frac{1}{4}} = k\sin\phi_{\frac{1}{4}}; \\[1em] \cdots \cdots \cdots \cdots \cdots \cdots \cdots \cdots \\[1em] \sin\phi_{\frac{1}{2^n}} = \dfrac{\sin\frac{1}{2}\phi_{\frac{1}{2^{n-1}}}}{\cos\frac{1}{2}\gamma_{\frac{1}{2^{n-1}}}}; \end{cases}$$

whence

$$(16)\quad E\phi = 2^n E\phi_{\frac{1}{2^n}} - \left(\sin\phi\sin^2\gamma_{\frac{1}{2}} + 2\sin\phi_{\frac{1}{2}}\sin^2\gamma_{\frac{1}{4}}\right.$$

$$\left. + 2^2\sin\phi_{\frac{1}{4}}\sin^2\gamma_{\frac{1}{8}} + \ldots 2^{n-1}\sin\phi_{\frac{1}{2^n}}\sin^2\gamma_{\frac{1}{2^{n-1}}}\right)$$

To find the limiting value, $E\phi_{\frac{1}{2^n}}$, we have, by the Binomial Theorem, since $\sin\phi = 1 - \dfrac{\phi^3}{\lfloor 3} + \dfrac{\phi^5}{\lfloor 5} - $ etc.,

$$\Delta\phi = (1 - k^2\sin^2\phi)^{\frac{1}{2}}$$

$$= 1 - \frac{k^2}{2}\left(\phi - \frac{\phi^3}{6}\right)^2 - \frac{k^4}{8}\left(\phi - \frac{\phi^3}{6}\right)^4 + \ldots$$

$$= 1 - \frac{k^2}{2}\phi^2 + \left(\frac{k^2}{6} - \frac{k^4}{8}\right)\phi^4.$$

Whence

$$Ek\phi_{\frac{1}{2^n}} = \int_0^{\phi_n}\Delta\phi_{\frac{1}{2^n}}d\phi$$

$$(17)\qquad = \phi_{\frac{1}{2^n}} - \frac{k^2}{6}\phi^3_{\frac{1}{2^n}} + \frac{k^2(4-3k^2)}{120}\phi^5_{\frac{1}{2^n}}.$$

Substituting in eq. (16) the numerical values derived from equations (15) and (17), we are enabled to determine the value of $E\phi$.

Landen's Transformation can also be applied to Elliptic Integrals of this class.

From eq. (11), Chap. IV, we get, by easy transformation,

(18) $\qquad \sin^2 2\phi = \sin^2 \phi_1 (1 + k_0 + 2k_0 \cos 2\phi)$.

From this we easily get

$$2k_0 \cos 2\phi \sin^2 \phi_1 = \sin^2 2\phi - \sin^2 \phi_1 - k_0^2 \sin^2 \phi_1$$
$$= 1 - \cos^2 2\phi - \sin^2 \phi_1 - k_0^2 \sin^2 \phi_1$$
$$= \Delta^2 k_0 \phi_1 - \sin^2 \phi_1 - \cos^2 2\phi\,;$$

whence

$$\cos^2 2\phi + 2k_0 \sin^2 \phi_1 \cos 2\phi = \Delta^2 k_0 \phi_1 - \sin^2 \phi_1\,;$$

and from this,

$$\cos 2\phi = -k_0 \sin^2 \phi_1 \pm \sqrt{\Delta^2 k_0 \phi_1 - \sin^2 \phi_1 + k_0^2 \sin^4 \phi_1}$$
(19) $\qquad = \cos \phi_1 \Delta k_0 \phi_1 - k_0 \sin^2 \phi_1\,;$

whence, also,

$$1 - \cos^2 2\phi = 1 - \cos^2 \phi_1 \Delta^2 \phi_1 + 2k \sin^2 \phi_1 \cos \phi_1 \Delta k_0 \phi_1 - k_0^2 \sin^4 \phi_1$$
$$= \sin^2 \phi_1 (1 + k_0^2 \cos^2 \phi_1 + 2k_0 \cos \phi_1 \Delta k_0 \phi_1 - k_0^2 \sin^2 \phi_1)$$

and

(20) $\qquad \sin 2\phi = \sin \phi_1 (\Delta k_0 \phi_1 + k_0 \cos \phi_1)$.

Differentiating equation (19), we get

$$2 \sin 2\phi \frac{d\phi}{d\phi_1} = \sin \phi_1 \frac{(k_0 \cos \phi_1 + \Delta k_0 \phi_1)^2}{\Delta k_0 \phi_1}.$$

Dividing this by equation (20), we have

$$\frac{2d\phi}{d\phi_1} = \frac{k_0 \cos \phi_1 + \Delta k_0 \phi_1}{\Delta k_0 \phi_1}.$$

But from (19), and eq. (6), Chap. IV,

$$k^2 \sin^2 \phi = \frac{k^2(1 - \cos 2\phi)}{2}$$

$$= \frac{2k_0}{(1+k_0)^2}\{1 + k_0 \sin^2 \phi_1 - \cos \phi_1 \Delta k_0 \phi_1\};$$

whence

$$\Delta k\phi = \frac{\Delta k_0 \phi_1 + k_0 \cos \phi_1}{1 + k_0},$$

and

$$2\Delta k\phi \cdot \frac{d\phi}{d\phi_1} = \frac{(k_0 \cos \phi_1 + \Delta k_0 \phi_1)^2}{(1+k_0)\Delta k_0 \phi_1},$$

and

$$d\phi \Delta k\phi = \frac{d\phi_1}{\Delta k_0 \phi_1} \cdot \frac{(k_0 \cos \phi_1 + \Delta k_0 \phi_1)^2}{2(1+k_0)}.$$

This gives immediately, by integration,

$$Ek\phi = \frac{1}{2(1+k_0)} \int \frac{d\phi_1}{\Delta k_0 \phi_1}\{k_0 \cos \phi_1 + \Delta k_0 \phi_1\}^2$$

$$= \frac{1}{2(1+k_0)} \int \frac{d\phi_1}{\Delta k_0 \phi_1}\{2\Delta^2 k_0 \phi_1 + 2k_0 \cos \phi_1 \Delta k_0 \phi_1 - k_1'^2\}$$

(21) $$= \frac{Ek_0 \phi_1}{1+k_0} + \frac{k_0 \sin \phi_1}{1+k_0} - \tfrac{1}{2}(1-k_0)Fk_0\phi_1.$$

Thus the value of $Ek\phi$ is made to depend upon $Ek_0\phi_1$ (containing a smaller modulus and a larger amplitude), and upon the integral of the first class, $Fk_0\phi_1$; k_0, ϕ_1, etc., being determined by the formulæ (6) to (12) of Chap. IV.

By successive applications of equation (21), $Ek\phi$ may be made to depend ultimately upon $Ek_{0n}\phi_n$, where k_{0n} approximates to zero and $Ek_{0n}\phi_n$ to ϕ_n.

Or, by reversing, it may be made to depend upon $Ek_n\phi_{0n}$, where k_n approximates to unity and $Ek_n\phi_{0n}$ to $-\cos\phi_{0n}$.

ELLIPTIC INTEGRALS OF THE SECOND ORDER.   87

To facilitate this, assume

$$Gk\phi = Ek\phi - Fk\phi.$$

Subtracting from equation (21) the equation

$$Fk\phi = \frac{1+k_0}{2} Fk_0\phi_1 \text{ (see eq. (13), Chap. IV)},$$

we have

$$Gk\phi = \frac{1}{1+k_0}(Gk_0\phi_1 + k_0 \sin \phi_1 - k_0 Fk_0\phi_1).$$

Repeated applications of this give

$$Gk_0\phi_1 = \frac{1}{1+k_{00}}(Gk_{00}\phi_2 + k_{00} \sin \phi_2 - k_{00} Fk_{00}\phi_2),$$

. . . .

$$Gk_{0(n-1)}\phi_{n-1} = \frac{1}{1+k_{0n}}(Gk_{0n}\phi_n + k_{0n} \sin \phi_n - k_{0n} Fk_{0n}\phi_n).$$

Whence

(22) $$Gk\phi = \Sigma_n^1 \left\{ \frac{k_{0n}(\sin \phi_n - Fk_{0n}\phi_n)}{[\overset{1}{\underset{n}{\Pi}}](1+k_{0n})} \right\} + \frac{Gk_{0n}\phi_n}{[\overset{1}{\underset{n}{\Pi}}](1+k_{0n})}.$$

But since (compare eq. (13), Chap. IV)

$$Fk\phi = \frac{Fk_{0n}\phi_n [\overset{1}{\underset{n}{\Pi}}](1+k_{0n})}{2^n},$$

or

(23) $$\frac{Fk_{0n}\phi_n}{[\overset{1}{\underset{n}{\Pi}}](1+k_{0n})} = \frac{2^n Fk\phi}{[\overset{1}{\underset{n}{\Pi}}](1+k_{0n})^2};$$

88   ELLIPTIC FUNCTIONS.

and since, also, (compare eq. (6), Chap. IV,)

$$\frac{k^2_{0(n-1)}}{k_{on}} = \frac{2^2}{(1 + k_{on})^2},$$

we have

(24) $\quad \dfrac{2^n k_{on}}{[\prod\limits_n^1](1 + k_{on})^2} = \dfrac{k_{on}}{2^n}\left[\prod\limits_n^1\right]\dfrac{k^2_{0(n-1)}}{k_{on}}$

$$= \frac{k_{on}}{2^n}\left[\prod_n^1\right]\frac{k_{0(n-1)}}{k_{on}}\left[\prod_n^1\right]k_{0(n-1)}$$

$$= \frac{k_{on}}{2^n} \cdot \frac{k}{k_0} \cdot \frac{k_0}{k_{00}} \cdots \frac{k_{0(n-1)}}{k_{on}} \cdot k\left[\prod_n^2\right]k_{0(n-1)}$$

$$= \frac{k^2}{2^n}\left[\prod_n^2\right]k_{0(n-1)}.$$

Substituting these values in equation (22), and neglecting the term containing $Gk_{on}\phi_n$ since, carried to its limiting value,

$$Gk_{on}\phi_n = Ek_{on}\phi_n - Fk_{on}\phi_n$$
$$= \phi_n - \phi_n = 0, \qquad (n = \text{limiting value,})$$

we have

(25) $\quad Gk\phi = \Sigma_n^1 \left\{ \dfrac{k\sqrt{k_{on}}\sin\phi_n \left[\prod\limits_n^2\right]\sqrt{k_{0(n-1)}} - k^2\left[\prod\limits_n^2\right]k_{0(n-1)}}{2^n} \right\}$

$$= k\left[\frac{\sqrt{k_0}}{2}\sin\phi_1 + \frac{\sqrt{k_0 k_{00}}}{2^2}\sin\phi_2 + \frac{\sqrt{k_0 k_{00} k_{03}}}{2^3}\sin\phi_3 + \cdots\right]$$

$$- \frac{k^2}{2}\left[1 + \frac{k_0}{2} + \frac{k_0 k_{00}}{2^2} + \frac{k_0 k_{00} k_{03}}{2^3} + \cdots\right];$$

whence

(26) $\quad Ek\phi = Fk\phi\left[1 - \dfrac{k^2}{2}\left(1 + \dfrac{k_0}{2} + \dfrac{k_0 k_{00}}{2^2} + \cdots\right)\right]$

$$+ k\left[\frac{\sqrt{k_0}}{2}\sin\phi_1 + \frac{\sqrt{k_0 k_{00}}}{2^2}\sin\phi_2 + \frac{\sqrt{k_0 k_{00} k_{03}}}{2^3}\sin\phi_3 + \cdots\right].$$

ELLIPTIC INTEGRALS OF THE SECOND ORDER. 89

From eq. (3), Chap. V, we see that when $\phi = \frac{\pi}{2}$,

$$\phi_n = 2^{n-1}\pi.$$

Substituting these values in equation (26), we have for a complete Elliptic Integral of the second class,

(27) $E\left(k, \frac{\pi}{2}\right) =$
$$F\left(k, \frac{\pi}{2}\right)\left[1 - \frac{k^2}{2}\left(1 + \frac{k_0}{2} + \frac{k_0 k_{00}}{2^2} + \frac{k_0 k_{00} k_{03}}{2^3} + \cdots\right)\right].$$

In a similar manner we could have found the formula for $E(k, \phi)$ in terms of an increasing modulus, viz.,

(28) $E(k, \phi) = F(k, \phi)\left[1 + k\left(1 + \frac{2}{k_1} + \frac{2^2}{k_1 k_2} + \frac{2^3}{k_1 k_2 k_3} + \cdots\right.\right.$
$$\left.\left. + \frac{2^{n-2}}{k_1 k_2 \cdots k_{n-2}} - \frac{2^{n-1}}{k_1 k_2 \cdots k_{n-1}}\right)\right]$$
$$- k\left[\sin \phi + \frac{2}{\sqrt{k}} \sin \phi_1 + \frac{2^2}{\sqrt{kk_1}} \sin \phi_2 + \cdots\right.$$
$$\left. + \frac{2^{n-1}}{\sqrt{kk_1 \cdots k_{n-2}}} \sin \phi_{n-1} - \frac{2^n}{\sqrt{kk_1 \cdots k_{n-1}}} \sin \phi_n\right].$$

CHAPTER XI.

ELLIPTIC INTEGRALS OF THE THIRD ORDER.

The Elliptic Integral of the third order is

(1) $$\Pi(n, k, \phi) = \int_0^\phi \frac{d\phi}{(1 + n \sin^2 \phi) \Delta\phi}.$$

Put

(2) $$\Pi(\phi) + \Pi(\psi) = S;$$

whence we have immediately

(3) $$dS = \frac{d\phi}{(1 + n \sin^2 \phi)\Delta\phi} + \frac{d\psi}{(1 + n \sin^2 \psi)\Delta\psi}.$$

But, eq. (2), Chap. II,

(4) $$\frac{d\phi}{\Delta\phi} + \frac{d\psi}{\Delta\psi} = 0;$$

whence

$$dS = \left(\frac{1}{1 + n \sin^2 \phi} - \frac{1}{1 + n \sin^2 \psi}\right)\frac{d\phi}{\Delta\phi}$$

(5) $$= \frac{n(\sin^2 \psi - \sin^2 \phi)}{(1 + n \sin^2 \phi)(1 + n \sin^2 \psi)} \cdot \frac{d\phi}{\Delta\phi}.$$

From equation (8), Chap. X, we get by differentiation, since σ (or μ) is constant,

$$\Delta\phi \cdot d\phi + \Delta\psi \cdot d\psi = k^2 \sin \sigma \, d(\sin \phi \sin \psi),$$

or, from equation (3),

$$(\sin^2 \psi - \sin^2 \phi) \frac{d\phi}{\Delta\phi} = \sin \sigma \, d(\sin \phi \sin \psi).$$

This, introduced into equation (5), gives

$$dS = \frac{n \sin \sigma \, d(\sin \phi \sin \psi)}{1 + n (\sin^2 \phi + \sin^2 \psi) + n^2 \sin^2 \phi \sin^2 \psi}.$$

Put
$$\sin \phi \sin \psi = q, \quad \sin^2 \phi + \sin^2 \psi = p;$$
whence

(6) $$dS = \frac{n \sin \sigma \, dq}{1 + np + n^2 p^2}.$$

From equation (5), Chap. II, we have

$$\cos \sigma = \cos \phi \cos \psi - \sin \phi \sin \psi \Delta \sigma,$$

from which we easily get

$$(\cos \sigma + q\Delta\sigma)^2 = \cos^2 \phi \cos^2 \psi$$
$$= (1 - \sin^2 \phi)(1 - \sin^2 \psi)$$
$$= 1 - p + q^2,$$

and thence
$$p = 1 + q^2 - (\cos \sigma + q\Delta\sigma)^2$$
$$= \sin^2 \sigma - 2 \cos \sigma \Delta \sigma q + k^2 \sin^2 \sigma \cdot q^2.$$

This, substituted in eq. (6), gives

$$dS = \frac{n \sin \sigma \, dq}{1 + n \sin^2 \sigma - 2n \cos \sigma \Delta \sigma q + n(n + k^2 \sin^2 \sigma)q^2}$$
$$= \frac{n \sin \sigma \, dq}{A - 2Bq + Cq^2},$$

where
$$A = 1 + n \sin^2 \sigma,$$
$$B = n \cos \sigma \Delta \sigma,$$
$$C = nk^2 \sin^2 \sigma + n^2.$$

From this we get

$$S = n \sin \sigma \int \frac{dq}{A - 2Bq + Cq^2} + \text{Const.}$$

In order to determine the constant of integration we must observe that for $\phi = 0$, $\psi = \sigma$ and $q = 0$; whence

$$\Pi\sigma = n \sin \sigma \int_{q=0} \frac{dq}{A - 2Bq + Cq^2} + \text{Const.};$$

whence

$$S = \Pi\sigma + n \sin \sigma \int_0^q \frac{dq}{A - 2Bq + Cq^2},$$

or

(7) $\quad \Pi\phi + \Pi\psi = \Pi\sigma + n \sin \sigma \int_0^q \frac{dq}{A - 2Bq + Cq^2}.$

But we have

$$dS = \frac{CM\,dq}{AC - B^2 + (Cq - B)^2}$$

$$= \frac{CM}{AC - B^2} \cdot \frac{dq}{1 + \left(\frac{Cq - B}{\sqrt{AC - B^2}}\right)^2}$$

$$= \frac{M}{\sqrt{AC - B^2}} \cdot \frac{\frac{C\,dq}{\sqrt{AC - B^2}}}{1 + \left(\frac{Cq - B}{\sqrt{AC - B^2}}\right)^2}$$

where $M = n \sin \sigma$.

The integral of the second member is

$$\frac{M}{\sqrt{AC - B^2}} \tan^{-1} \frac{Cq - B}{\sqrt{AC - B^2}};$$

whence

$$\int_0^q dS = S_1 = \frac{M}{\sqrt{AC-B^2}}\left[\tan^{-1}\frac{Cq-B}{\sqrt{AC-B^2}}+\tan^{-1}\frac{B}{\sqrt{AC-B^2}}\right];$$

or, since

$$\tan^{-1} x + \tan^{-1} y = \tan^{-1}\frac{x+y}{1-xy},$$

$$S_1 = \frac{M}{\sqrt{AC-B^2}} \tan^{-1}\frac{q\sqrt{AC-B^2}}{A-Bq}.$$

Substituting the values of A, B, C and M, we have

$$AC - B^2 = n(1 + n - \Delta^2\sigma)(1 + n \sin^2 \sigma) - n^2 \cos^2 \sigma \Delta^2\sigma$$
$$= n(1 + n - \Delta^2\sigma + n(1 + n) \sin^2 \sigma - n\Delta^2\sigma)$$
$$= n(1 + n)(1 - \Delta^2\sigma + n \sin^2 \sigma)$$
$$= n(1 + n)(k^2 + n) \sin^2 \sigma;$$

and putting

$$\frac{(1+n)(k^2+n)}{n} = \Omega,$$

we have

$$\sqrt{AC-B^2} = n\sqrt{\Omega} \sin \sigma.$$

Substituting these values in eq. (7), we have

$$\Pi(n, k, \phi) + \Pi(n, k, \psi) - \Pi(n, k, \sigma) = S_1$$
$$= \frac{1}{\sqrt{\Omega}} \tan^{-1}\frac{n\sqrt{\Omega} \sin \phi \sin \psi \sin \sigma}{1+n \sin^2 \sigma - n \sin \phi \sin \psi \cos \sigma \Delta \sigma}.$$

CHAPTER XII.

NUMERICAL CALCULATIONS. q.

CALCULATION OF THE VALUE OF q.

FROM eq. (7), Chap. IX, we have

$$\operatorname{dn} u = \sqrt{k'}\, \frac{\Theta(u+K)}{\Theta(u)};$$

whence, eq. (9), Chap. IV, eqs. (27) and (39), Chap. IX,

(1) $$\sqrt{\cos \theta} = \frac{1 - 2q + 2q^4 - 2q^9 + 2q^{16} - \ldots}{1 + 2q + 2q^4 + 2q^9 + 2q^{16} + \ldots}$$

$$= 1 - 4q + 8q^2 - 16q^3 + 32q^4 - 56q^5 + \ldots$$

The first five terms of this series can be represented by

$$\sqrt{\cos \theta} = \frac{1 - 2q}{1 + 2q}.$$

From this we get

(2) $$q = \frac{1}{2} \cdot \frac{1 - \sqrt{\cos \theta}}{1 + \sqrt{\cos \theta}},$$

which is exact up to the term containing q^5.

Or we can deduce a more exact formula as follows: From eq. (1),

$$\frac{1 + \sqrt{\cos \theta}}{1 - \sqrt{\cos \theta}} = \frac{\sqrt{1 + \tan^2 \tfrac{1}{2}\theta} + \sqrt{1 - \tan^2 \tfrac{1}{2}\theta}}{\sqrt{1 + \tan^2 \tfrac{1}{2}\theta} - \sqrt{1 - \tan^2 \tfrac{1}{2}\theta}}$$

$$= \frac{1 + 2q^4 + 2q^{16} + \ldots}{2q + 2q^9 + 2q^{25} + \ldots};$$

whence, by the method of indeterminate coefficients,

(3) $\quad q = \tfrac{1}{4}\tan^2\tfrac{\theta}{2} + \tfrac{1}{16}\tan^6\tfrac{\theta}{2} + \tfrac{67}{512}\tan^{10}\tfrac{\theta}{2} + \tfrac{45}{2048}\tan^{14}\tfrac{\theta}{2} + \ldots,$

or

$\log q = 2\log\tan\tfrac{\theta}{2} - \log 4 + $

$\qquad \log(1 + \tfrac{1}{4}\tan^4\tfrac{\theta}{2} + \tfrac{17}{128}\tan^8\tfrac{\theta}{2} + \tfrac{45}{512}\tan^{12}\tfrac{\theta}{2} \ldots)$

(4) $\quad = 2\log\tan\tfrac{\theta}{2} - \log 4 + $

$\qquad M(\tfrac{1}{4}\tan^4\tfrac{\theta}{2} + \tfrac{13}{128}\tan^8\tfrac{\theta}{2} + \tfrac{23}{384}\tan^{12}\tfrac{\theta}{2} + \ldots),$

M being the modulus of the common system of logarithms. Put

(5) $\quad \log q = 2\log\tan\tfrac{\theta}{2} + 9.397940 + a\tan^4\tfrac{\theta}{2} + $

$\qquad\qquad b\tan^8\tfrac{\theta}{2} + c\tan^{12}\tfrac{\theta}{2} + \ldots,$

in which

$\qquad \log a = 9.0357243;$
$\qquad \log b = 8.64452;$
$\qquad \log c = 8.41518;$
$\qquad \log d = 8.25283.$

EXAMPLE. Let $k' = \cos 10° 23' 46''$. To find q.

$4\log\tan\tfrac{\theta}{2} = 5.835 \qquad 2\log\tan\tfrac{\theta}{2} = 7.9176842$

$\log a = 9.036 \qquad\qquad\qquad 9.3979400$

$\overline{4.871} \qquad\qquad\qquad\qquad 74$

$\qquad\qquad\qquad\qquad\qquad \overline{\log q = 7.3156316}$

$a\tan^4\tfrac{\theta}{2} = 0.0000074$

When θ approaches 90°, $\tan\dfrac{\theta}{2}$ differs little from unity, and the series in eq. (5) is not very converging, but q can be calculated by means of eq. (6), Chap. VII, viz.,

$$q = e^{-\frac{\pi K'}{K}}, \qquad q' = e^{-\frac{\pi K}{K'}}.$$

By comparing these equations with eqs. (6) and (9), Chap. IV, we see that if
then
$$q = f(k) = f(\theta),$$
$$q' = f(k') = f(90° - \theta).$$

Therefore, having θ, we can from its complement, $90° - \theta$, find q' by eq. (5), and thence q by the following process. We have

$$\frac{1}{q} = e^{\frac{\pi K'}{K}}, \qquad \frac{1}{q'} = e^{\frac{\pi K}{K'}};$$

whence

$$\log\frac{1}{q}\log\frac{1}{q'} = M'^2\pi^2 = 1.8615228,$$

(6) $\qquad \log\log\dfrac{1}{q} + \log\log\dfrac{1}{q'} = 0.2698684,$

by which we can deduce q from q'.

EXAMPLE. Let $\theta = 79° \, 36' \, 14''$. To find q.

$$90° - \theta = 10° \, 23' \, 46''.$$

By eq. (5) we get

$$\log q' = 7.3156316, \qquad \log\frac{1}{q'} = 2.6843684,$$

$$\text{and} \quad \log\log\frac{1}{q'} = .4288421\,;$$

and by eq. (6),
$$\log \log \frac{1}{q} = 9.8410263;$$
whence
$$\log q = \bar{1}.3065321.$$

When $k' = k = \cos 45° = \frac{1}{2}\sqrt{2}$, eq. (6) becomes

(7) $$\log \frac{1}{q} = M\pi = 1.3643763; \qquad (k = k';)$$
whence
$$\log q = \bar{2}.6356237,$$
$$q = 0.0432138. \qquad (k = k'.)$$

EXAMPLE. Given $\theta = 10° 23' 46''$. Find q.
$$\textit{Ans. } \log q = 7.3156316.$$

EXAMPLE. Given $\theta = 82° 45'$. Find q.
$$\textit{Ans. } \log q = 9.37919.$$

CHAPTER XIII.

NUMERICAL CALCULATIONS. K.

CALCULATION OF THE VALUE OF K.

WE have already found from eq. (37), Chap. IX,

(1) $$\Theta(0) = \sqrt{\frac{2k'K}{\pi}},$$

and from eq. (40), same chapter,

(2) $$\Theta(K) = \frac{\Theta(0)}{\sqrt{k'}} = \sqrt{\frac{2K}{\pi}}.$$

But, eqs. (38) and (27), Chap. IX,

$$\Theta(K) = 1 + 2q + 2q^4 + 2q^9 + 2q^{16} + \cdots,$$

$$\Theta(0) = 1 - 2q + 2q^4 - 2q^9 + 2q^{16} - \cdots;$$

whence, eq. (2),

(3) $$K = \frac{\pi}{2}(1 + 2q + 2q^4 + 2q^9 + \cdots)^2.$$

By adding eqs. (1) and (2) we get

$$\Theta(0) + \Theta(K) = \sqrt{\frac{2K}{\pi}}(1 + \sqrt{k'});$$

whence

$$K = \frac{\pi}{2}\left(\frac{\Theta(0) + \Theta(K)}{1 + \sqrt{k'}}\right)^2$$

$$= \frac{\pi}{2}\left[\frac{2(1 + 2q^4 + 2q^{16} + \cdots)}{1 + \sqrt{k'}}\right]^2.$$

(4) $$= \frac{\pi}{2}\left(\frac{2}{1+\sqrt{k'}}\right)^2 (1 + 2q' + 2q'^4 + \ldots)^2.$$

EXAMPLE. Let $k = \sin\theta = \sin 19°\,30'$. Required K.
First Method. By eq. (3).

By eq. (5), Chap. XII, we find $\log q = 8.6356236$. Applying eq. (3), using only two terms of the series, we have

$$1 + 2q = 1.0147662$$
$$\log(1 + 2q) = 0.0063660$$
$$2\log(1 + 2q) = 0.0127320$$
$$\log \frac{\pi}{2} = 0.1961199$$
$$\overline{\log K = 0.2088519}$$
$$K = 1.615101$$

Second Method. By eq. (4).
Equation (4) may be written, neglecting q^4,

$$K = \frac{\pi}{2}\left(\frac{1+\sqrt{\cos\theta}}{2}\right)^{-2};$$

whence

$$\log\cos\theta = 9.9743466,$$
$$\log\sqrt{\cos\theta} = 9.9871733,$$
$$1 + \sqrt{\cos\theta} = 1.9708973,$$
$$\frac{1+\sqrt{\cos\theta}}{2} = 0.98544865;$$

and

$$\log K = 0.2088519,$$
$$K = 1.615101,$$

the same result as above.

ELLIPTIC FUNCTIONS.

Third Method. By eq. (7), Chap. V.

$$\theta = 19° 30'$$
$$\tfrac{1}{2}\theta = 9° 45'$$
$$\log \tan \tfrac{1}{2}\theta = 9.235103$$
$$\log \cos \tfrac{1}{2}\theta = 9.993681$$
$$\left.\begin{array}{l}\log \tan^2 \tfrac{1}{2}\theta \\ \log \sin \theta_0\end{array}\right\} = 8.470206$$
$$\theta_0 = 1° 41' 31''.1$$

$$\theta_0 = 1° 41' 31''.1$$
$$\tfrac{1}{2}\theta_0 = 0° 50' 45''.5$$

$$\log \cos \tfrac{1}{2}\theta_0 = 9.999953$$

$$\log \cos^2 \tfrac{1}{2}\theta = 9.987362$$
$$\log \cos^2 \tfrac{1}{2}\theta_0 = 9.999906$$
$$\overline{9.987268}$$
$$\log \tfrac{\pi}{2} = 0.196120$$
$$\overline{\log K = 0.208852}$$

θ_{00} is not calculated, as it is evident that its cosine will be 1.
EXAMPLE. Given $k = \sin 75°$. Find K.
By eq. (7), Chap. V.
From eqs. (14$_1$), Chap. IV, we find

$k = \sin \theta = \sin 75°$ $\qquad$ $\log = 9.9849438$

$$k_0 = \begin{cases} \tan^2 \tfrac{1}{2}\theta = \tan^2 37° 30' \\ \sin \theta_0 = \sin 36° 4' 16''.47 \end{cases} \qquad 9.7699610$$

$$k_{00} = \begin{cases} \tan^2 \tfrac{1}{2}\theta_0 = \tan^2 18° 2' 8''.235 \\ \sin \theta_{00} = \sin 6° 5' 9''.38 \end{cases} \qquad 9.0253880$$

$$k_{03} = \begin{cases} \tan^2 \tfrac{1}{2}\theta_{00} = \tan^2 3° 2' 34''.69 \\ \sin \theta_{03} = \sin 9' 42''.90 \end{cases} \qquad 7.4511672$$

NUMERICAL CALCULATIONS. K.

		log	2 log	a. c. 2 log
$\cos \tfrac{1}{2}\theta$	$= \cos 37° 30'$	9.8994667	9.7989334	0.2010666
$\cos \tfrac{1}{2}\theta_0$	$= \cos 18° 2'.13725$	9.9781184	9.9562368	0.0437632
$\cos \tfrac{1}{2}\theta_{02}$	$= \cos 3° 2'.57817$	9.9993873	9.9987746	0.0012254
$\cos \tfrac{1}{2}\theta_{03}$	$= \cos 4'.8575$	9.9999995	9.9999990	0.0000010

$$ 0.2460562$$

$\dfrac{\pi}{2}$ $\dfrac{\pi}{2}$ .1961199

$$\log K = 0.4421761$$
$$K = 2.768064 \quad Ans.$$

EXAMPLE. Given $k = \sin 45°$. Find K.
Method of eq. (7), Chap. V.
From eqs. (14$_1$), Chap. IV, we have

 log

$k_0 = \begin{cases} \tan^2 \tfrac{1}{2}\theta &= \tan^2 22° 30' \\ \sin \theta_0 &= \sin 9° 52'.75683 \end{cases}$ 9.2344486

$k_{00} = \begin{cases} \tan^2 \tfrac{1}{2}\theta_0 &= \tan^2 4° 56'.37841 \\ \sin \theta_{00} &= \sin 25'.679 \end{cases}$ 7.8733009

$k_{03} = \begin{cases} \tan^2 \tfrac{1}{2}\theta_{00} &= \tan^2 12'.3395 \\ \sin \theta_{03} &= \sin 0'.05 \end{cases}$ 5.1445523

 a. c. log $\cos^2 \tfrac{1}{2}\theta$ 0.0687694
 a. c. log $\cos^2 \tfrac{1}{2}\theta_0$ 0.0032320
 a. c. log $\cos^2 \tfrac{1}{2}\theta_{00}$ 0.0000060

 $\log \dfrac{\pi}{2}$ 0.1961199

$$\log K = 0.2681273$$
$$K = 1.8540747 \quad Ans.$$

EXAMPLE. Given $\theta = 63° 30'$. Find K.
 *Ans.* $\log K = 0.3539686$.

EXAMPLE. Given $\theta = 34° 30'$. Find K.
 *Ans.* $K = 1.72627$.

CHAPTER XIV.

NUMERICAL CALCULATIONS. *u.*

CALCULATION OF THE VALUE OF *u*.

WHEN $\theta^\circ = \sin^{-1} k < 45^\circ$.

EXAMPLE. Let $\phi = 30^\circ$, $k = \sin 45^\circ$. Find u.

First Method. Eq. (23), Chap. IV, and eqs. (14_1), (14_2), (14_3), Chap. IV.

By equations (14_1),

$$\frac{\theta}{2} = 22^\circ \, 30';$$

$$\log \tan \frac{\theta}{2} = 9.6172243;$$

$$\log \tan^2 \frac{\theta}{2} = 9.2344486 = \log k_0 = \log \sin \theta_0;$$

$$\theta_0 = 9^\circ \, 52' \, 45''.41;$$

$$\log \tan \frac{\theta_0}{2} = 8.9366506;$$

$$\log \tan^2 \frac{\theta_0}{2} = 7.8733012 = \log k_{00} = \log \sin \theta_{00};$$

$$\theta_{00} = 0^\circ \, 25' \, 40''.7;$$

$$\log \tan^2 \frac{\theta_{00}}{2} = 5.144552 = \log k_{0s}.$$

By equations (14$_2$),

$$\phi = 30°$$
$$\log \tan \phi = 9.761439$$
$$\underline{\log \cos \theta = 9.849485}$$
$$\log \tan (\phi_1 - \phi) = 9.610924$$
$$\phi_1 - \phi = 22° \; 12' \; 27''.56$$
$$\phi_1 = 52° \; 12' \; 27''.56$$

$$\log \tan \phi_1 = 0.110438$$
$$\underline{\log \cos \theta_0 = 9.993512}$$
$$\log \tan (\phi_2 - \phi_1) = 0.103949$$
$$\phi_2 - \phi_1 = 51° \; 47' \; 32''.59$$

$$\phi_2 = 104° \; 0' \; 0''.15$$
$$\log \tan \phi_2 = 0.603228$$
$$\underline{\log \cos \theta_{00} = 9.999988}$$
$$\log \tan (\phi_3 - \phi_2) = 0.603216$$
$$\phi_3 - \phi_2 = 104° \; 0' \; 1''.5$$

$$\phi_3 = 208° \; 0' \; 1''.65$$

Since $\dfrac{\phi_2}{4} = 26° \; 0' \; 0''.04$ and $\dfrac{\phi_3}{8} = 26° \; 0' \; 0''.21$,

we need not calculate ϕ_4.

$$\frac{\phi_3}{8} = 93600''.21.$$

Reducing this to radians, we have

$$\log \frac{\phi_3}{8} = 9.656852.$$

Substituting in eq. (23), Chap. IV, we have, since $\cos \theta_{03} = 1$,

a. c. $\log \cos \theta = 0.150515$
$\log \cos \theta_0 = 9.993512$
$\log \cos \theta_{00} = 9.999988$
$\overline{ 0.144014}$

$0.072007 = \log \sqrt{\dfrac{\cos \theta_0 \cos \theta_{00}}{\cos \theta}}$

$\log \dfrac{\phi_3}{8} = 9.656852$
$\overline{\log u = 9.728859}$
$u = 0.535623,$ *Ans.*

When $\theta = \sin^{-1} k > 45°$.

EXAMPLE. Given $k = \sin 75°$, $\tan \phi = \sqrt{\dfrac{2}{\sqrt{3}}}$. To find $F(k, \phi)$.

First Method.   Bisected Amplitudes.

By equations (24) and (25), Chap. IV, we get

$\phi\ \ = 47°\ \ 3'\ 30''.91,$
$\phi_{\frac{1}{2}} = 25°\ 36'\ \ 5''.64,\qquad \beta\ \ = 45°;$
$\phi_{\frac{1}{4}} = 13°\ \ 6'\ 30''.98,\qquad \beta_0 = 24°\ 40'\ 10''.94;$
$\phi_{\frac{1}{8}} = 6°\ 35'\ 40''.74,\qquad \beta_{00} = 12°\ 39'\ 15''.83;$
$\phi_{\frac{1}{16}} = 3°\ 18'\ \ 8''.75,\qquad \beta_{03} = 6°\ 22'\ \ 8''.40;$
$\phi_{\frac{1}{32}} = 1°\ 39'\ \ 7''.43,\qquad \beta_{04} =$

Substituting in equation (26), Chap. IV, we have

$F(k, \phi) = 32 \times 1°\ 39'\ 7''.43$
$ = 52°\ 51'\ 58''.03$
$ = 0.9226878.$ *Ans.*

Second Method. Equation (29), Chap. IV.

From equations (18$_2$), Chap. IV, we have

$$
\begin{array}{llll}
 & & & \log \\
k = \cos \eta & = \cos 15° \ 0' \ 0''.00 & & 9.9849438 \\
k' = \sin \eta & = \sin 15° \ 0' \ 0''.00 & & 9.4129962 \\
k_0' = \begin{cases} \tan^2 \tfrac{1}{2}\eta = \tan^2 7° \ 30' \ 0''.00 \\ \sin \eta_0 = \sin 0° \ 59' \ 35''.25 \end{cases} & & & 8.2388582 \\
k_1 = \cos \eta_0 & = \cos 0° \ 59' \ 35''.25 & & 9.9999348 \\
k_{00}' = \begin{cases} \tan^2 \tfrac{1}{2}\eta_0 = \tan^2 0° \ 29' \ 47''.62 \\ \sin \eta_{00} = \sin 0° \ 0' \ 15''.49 \end{cases} & & & 5.8757219 \\
k_2 = \cos \eta_{00} & = \cos 0° \ 0' \ 15''.49 & & 0.0000000 \\
k_{03}' = (\tfrac{1}{2} k_{00}')^2 & & & 1.1493838 \\
\end{array}
$$

From equations (18$_2$), Chap. IV, we get

$$\phi = 47° \ 3' \ 30''.95;$$
$$2\phi_0 - \phi = 45°;$$
$$\phi_0 = 46° \ 1' \ 45''.475;$$
$$\phi_{02} = 46° \ 1' \ 29''.41;$$
$$\phi_{03} = 46° \ 1' \ 29''.41;$$
$$45° + \tfrac{1}{2}\phi_3 = 68° \ 0' \ 44''.705.$$

Substituting these values in eq. (29), Chap. IV, we get

$$F(k, \phi) = \sqrt{\frac{k_1}{k}} \cdot \frac{1}{M} \cdot \log \tan 68° \ 0' \ 44''.705$$

$$= 0.9226877. \quad Ans.$$

Third Method. Equation (23)*, Chap. IV.

From equations (14$_1$), Chap. IV, we have

$$k = \sin \theta \quad = \sin 75° \ 0' \ 0'' \quad \log = 9.9849438$$
$$k' = \cos \theta \quad = \cos 75° \quad\quad\quad\quad\quad\quad 9.4129962$$
$$k_0 = \begin{cases} \tan^2 \tfrac{1}{2} \theta = \tan^2 37° \ 30' \\ \sin \theta_0 = \sin 36° \ 4' \ 16''.47 \end{cases} \quad 9.7699610$$
$$k_1' = \cos \theta_0 \quad\quad\quad\quad\quad\quad\quad\quad\quad 9.9075648$$
$$k_{02} = \begin{cases} \tan^2 \tfrac{1}{2} \theta_0 = \tan^2 18° \ 2' \ 8''.235 \\ \sin \theta_{00} = \sin \ \ 6° \ 5' \ 9''.38 \end{cases} \quad 9.0253880$$
$$k_2' = \cos \theta_{00} \quad\quad\quad\quad\quad\quad\quad\quad\quad 9.9975452$$
$$k_{03} = \begin{cases} \tan^2 \tfrac{1}{2} \theta_{00} = \tan^2 \ 3° \ 2' \ 34''.69 \\ \sin \theta_{03} = \sin \quad\quad 9' \ 42''.90 \end{cases} \quad 7.4511672$$
$$k_3' = \cos \theta_{03} \quad\quad\quad\quad\quad\quad\quad\quad\quad 9.9999982$$
$$k_{04} = (\tfrac{1}{2} k_{03})^2 \quad\quad\quad\quad\quad\quad\quad\quad\quad 4.3002761$$
$$k_4' = \quad\quad\quad\quad\quad\quad\quad\quad\quad\quad\quad\quad 0.0000000$$

From equations (14$_2$), Chap. IV, we have

$$\phi = \ \ 47° \ \ 3' \ 30''.94;$$
$$\phi_1 = \ \ 62° \ 36' \ \ 3''.10;$$
$$\phi_2 = 119° \ 55' \ 47''.67;$$
$$\phi_3 = 240° \ \ 0' \ \ 0''.19;$$
$$\phi_4 = 480° \ \ 0' \ \ 0''.$$

Therefore the limit of ϕ, $\dfrac{\phi_1}{2}$, $\dfrac{\phi_2}{4}$, or $\dfrac{\phi_n}{2^n}$ is $30° = \dfrac{\pi}{6}$.

Substituting these values in eq. (23)*, Chap. IV, we have

$$F(k, \phi) = \sqrt{\frac{k_1' k_2' k_3' k_4'}{k'}} \cdot \frac{\pi}{6}$$

$$= 0.9226874. \quad Ans.$$

EXAMPLE. Given $\phi = 30°$, $k = \sin 89°$. Find u.
Method of eq. (28), Chap. IV.

From eqs. (18$_1$) we find

$$k_1 = \sin \theta_1 \quad \text{and} \quad \tan^2 \tfrac{1}{2} \theta_1 = k = \sin \theta,$$

from which we find that $k_1 = 1$ as far as seven decimal places. From eqs. (18$_2$) we have

$$\begin{aligned} \sin \phi &= 9.6989700 \\ k &= 9.9999338 \\ \hline \sin (2\phi_0 - \phi) &= 9.6989038 \\ 2\phi_0 - \phi &= 29° \ 59'.69733 \\ 2\phi_0 &= 59° \ 59'.69733 \\ 45° + \tfrac{1}{2} \phi_0{}^* &= 59° \ 59'.92433 \\ \log(45° + \tfrac{1}{2} \phi_0) &= 0.2385385 \end{aligned}$$

From eqs. (18$_3$), Chap. IV, we have

$$k = \cos \eta = \cos 1°, \qquad \tfrac{1}{2}\eta = 30'.$$

Substituting in eq. (28), Chap. IV, we have

$$\begin{aligned} \text{a. c. } \log \cos \tfrac{1}{2}\eta \quad & 0.0000330 \\ \log \log (45° + \tfrac{1}{2}\phi_0) \quad & 9.3775585 \\ \text{a. c. } \log M \quad & 0.3622157 \\ \hline \log F(k, \phi) &= 9.7398072 \\ F(k, \phi) &= 0.549297. \quad Ans. \end{aligned}$$

EXAMPLE. Given $\phi = 79°$, $k = 0.25882$. Find u.
Ans. $u = 0.39947$.

EXAMPLE. Given $\phi = 37°$, $k = 0.86603$. Find u.
Ans. $u = 0.68141$.

---

* Since $k_1 = 1$, $\phi_{00} = \phi_0$, and we need not carry the calculation further.

CHAPTER XV.

NUMERICAL CALCULATIONS. ϕ.

EXAMPLE. Given $u = 1.368407$, $\theta = 38°$. Find ϕ.
First Method. From eqs. (46) and (41)*, Chap. IX, we have

$$u = x\Theta^2(K),$$

$$\Delta\phi = \sqrt{k'}\,\frac{\Theta_1(x)}{\Theta(x)}.$$

From equations (5), Chap. XII, and (38), Chap. IX, we have

$$\log q = 8.4734187$$
$$\log \Theta^2(K) = 0.0501955$$
$$\log u = 0.1362153$$
$$\overline{\log x = 0.0860198}$$
$$x = 69°\ 50'\ 46''.12$$

From equations (23) and (24), Chap. IX, we get

$$\log \Theta_1(x) = 9.9798368$$
$$\log \Theta(x) = 0.0192687$$
$$\overline{9.9605681}$$
$$\log \sqrt{k'} = 9.9482661$$
$$\overline{\log \Delta\phi = 9.9088342 = \log \sin \lambda}$$

But

$$k^2 \sin^2 \phi = 1 - \Delta^2\phi,$$
$$k \sin \phi = \cos \lambda;$$

whence
$$\log \cos \lambda = 9.7675483$$
$$\log k = 9.7893420$$
$$\log \sin \phi = 9.9782063$$
$$\phi = 72°. \quad Ans.$$

Second Method. From eq. (1), Chap. VI.
From eqs. (14$_1$) Chap. IV, we find

$$k_0 = \begin{cases} \tan^2 \tfrac{1}{2}\theta = \tan^2 19° \\ \sin \theta_0 = \sin 6° 48'.54569 \\ \cos \theta_0 \end{cases} \log = 9.0739438$$
$$\qquad\qquad\qquad\qquad\qquad\quad 9.9969260$$

$$k_{00} = \begin{cases} \tan^2 \tfrac{1}{2}\theta_0 = \tan^2 3° 24'.2784 \\ \sin \theta_{00} = 12'.16659 \\ \cos \theta_{00} \end{cases} 7.5488952$$
$$\phantom{k_{00} =}\phantom{\cos \theta_{00}}\qquad\qquad\qquad\qquad\qquad 9.9999974$$

$$k_{03} = \begin{cases} \tan^2 \tfrac{1}{2}\theta_{00} = \tan^2 6'.08329 \\ \sin \theta_{03} \\ \cos \theta_{03} \end{cases} 4.4957316$$
$$\phantom{k_{03} =}\phantom{\cos \theta_{03}}\qquad\qquad\qquad\qquad\qquad 0.0000000$$

Substituting these values in eq. (1), Chap. VI, we have

$$\log \cos \theta_0 \quad 9.9969260$$
$$\log \cos \theta_{00} \quad 9.9999974$$
$$\overline{}$$
$$9.9969234$$
$$\log \sqrt{\cos \theta_0 \cos \theta_{00}} \quad 9.9984617$$
$$\text{a. c. log} \phantom{\sqrt{\cos \theta_0 \cos \theta_{00}}} \text{`` ``} \quad 0.0015383$$
$$\log u \quad .1362153$$
$$\log \sqrt{\cos \theta} \quad 9.9482660$$
$$\log 2^3 \quad .9030900^*$$
$$\text{a. c. log } \sqrt{\cos \theta_0 \cos \theta_{00}} \quad 0.0015383$$
$$\overline{}$$
$$0.9891096$$
$$\bar{2}.2418773$$
$$\log \phi_3^* \quad 2.7472323$$
$$\phi_3 \quad 558° 46'.140$$

---
* n is taken equal to 3, because $\cos_{03} = 1$.

ELLIPTIC FUNCTIONS.

Whence, by equations (1)* of Chap. VI, we get

$$\begin{aligned}
k_{02} \;\; \log &= 4.4957316 \\
\sin \phi_2 &\quad 9.5075232_n \\
\hline
\sin (2\phi_2 - \phi_3) &\quad 4.0032548_n \\
2\phi_2 - \phi_3 &= -0'.00346 \\
\phi_2 &= 279° \; 23'.06827
\end{aligned}$$

$$\begin{aligned}
k_{0u} \;\; \log &= 7.5488952 \\
\sin \phi_2 &\quad 9.9941484_n \\
\hline
\sin (2\phi_1 - \phi_2) &\quad 7.5430436_n \\
2\phi_1 - \phi_2 &= -12'.0039 \\
\phi_1 &= 139° \; 35'.5321
\end{aligned}$$

$$\begin{aligned}
k_0 \;\; \log &= 9.0739438 \\
\sin \phi_1 &\quad 9.8117249 \\
\hline
\sin (2\phi - \phi_1) &\quad 8.8856687 \\
2\phi - \phi_1 &= 4° \; 24'.467 \\
\phi &= 71° \; 59'.9999 \\
&= 72°. \;\; Ans.
\end{aligned}$$

EXAMPLE. Given $u = 2.41569$, $\theta = 80°$. Find ϕ.

Ans. $\phi = 82°$.

EXAMPLE. Given $u = 1.62530$, $k = \frac{1}{2}$. Find ϕ.

Ans. $\phi = 87°$.

CHAPTER XVI.

NUMERICAL CALCULATIONS. $E(k, \phi)$.

First Method. By Chap. X, eqs. (15), (16), and (17).
EXAMPLE. Given $k = 0.9327$, $\phi = 80°$. Find $E(k, \phi)$.
By eq. (15), Chap. X,

$\phi = 80°$; $\gamma = 67° 44'.$;
$\phi_{\frac{1}{2}} = 50° 43'.6,$ $\gamma_{\frac{1}{2}} = 46° 40'.4;$
$\phi_{\frac{1}{4}} = 27° 48'.5,$ $\gamma_{\frac{1}{4}} = 26° 0'.1;$
$\phi_{\frac{1}{8}} = 14° 16'.7,$ $\gamma_{\frac{1}{8}} = 13° 24'.0;$
$\phi_{\frac{1}{16}} = 7° 11'.3,$ $\gamma_{\frac{1}{16}} = 6° 45'.2;$
$\phi_{\frac{1}{32}} = 3° 36'.0,$ $\log \sin \gamma_{\frac{1}{32}} = 8.77094;$
$\phi_{\frac{1}{32}} = 0.062831.$
$\therefore \phi_{\frac{1}{32}}^{5} < 0.0000001.$

Whence, by eq. (17),

$$E(k, \phi_{\frac{1}{32}}) = 0.062794$$
$\sin \phi \quad \sin^2 \gamma_{\frac{1}{2}} = 0.52116$
$2 \sin \phi_{\frac{1}{2}} \quad \sin^2 \gamma_{\frac{1}{4}} = 0.29757$
$4 \sin \phi_{\frac{1}{4}} \quad \sin^2 \gamma_{\frac{1}{8}} = 0.10023$
$8 \sin \phi_{\frac{1}{8}} \quad \sin^2 \gamma_{\frac{1}{16}} = 0.02728$
$16 \sin \phi_{\frac{1}{16}} \quad \sin^2 \gamma_{\frac{1}{32}} = 0.00697$
$$\overline{\quad 0.95321 \quad}$$

Hence, by eq. (16),

$$E(k, \phi) = 32 E(k, \phi_{\frac{1}{32}}) - 0.95321$$
$$= 2.0094 - 0.9532 = 1.0562.$$

112   *ELLIPTIC FUNCTIONS.*

Second Method. By Chap. X, eq. (26).

EXAMPLE. Given $k = \sin 75°$, $\tan \phi = \sqrt{\dfrac{2}{\sqrt{3}}}$. Find $E(k, \phi)$.

From eqs. (14$_1$), Chap. IV, we have

$k = \sin \theta = \sin 75° \; 0' \; 0''$ \qquad $\log = 9.9849438$

$k' = \cos \theta = \cos 75°$ \qquad 9.4129962

$k_0 = \begin{cases} \tan^2 \tfrac{1}{2}\theta = \tan^2 37° \; 30' \\ \sin \theta_0 = \sin 36° \; 4' \; 16''.47 \end{cases}$ \qquad 9.7699610

$k_1' = \cos \theta_0$ \qquad 9.9075648

$k_{02} = \begin{cases} \tan^2 \tfrac{1}{2}\theta_0 = \tan^2 18° \; 2' \; 8''.235 \\ \sin \theta_{00} = \sin 6° \; 5' \; 9''.38 \end{cases}$ \qquad 9.0253880

$k_2' = \cos \theta_{00}$ \qquad 9.9975452

$k_{03} = \begin{cases} \tan^2 \tfrac{1}{2}\theta_{00} = \tan^2 3° \; 2' \; 34''.69 \\ \sin \theta_{03} = \sin 9' \; 42''.90 \end{cases}$ \qquad 7.4511672

$k_3' = \cos \theta_{03}$ \qquad 9.9999982

$k_{04} = (\tfrac{1}{2}k_{03})^2$ \qquad 4.3002761

$k_4' =$ \qquad 0.0000000

From eqs. (14$_2$), Chap. IV, we have

$\phi = 47° \; 3' \; 30''.94;$

$\phi_1 = 62° \; 36' \; 3''.10;$

$\phi_2 = 119° \; 55' \; 47''.67;$

$\phi_3 = 240° \; 0' \; 0''.19.$

Applying eq. (26), Chap. X, we have

$$k^2 \quad \log = 9.9698876$$
a. c. 2   9.6989700

	9.6688576	.4665064
k_0	9.7699610	
a. c. 2	9.6989700	
	9.1377886	.1373373
k_{00}	9.0253880	
a. c. 2	9.6989700	
	7.8621466	.0072802
k_{03}	7.4511672	
a. c. 2	9.6989700	
	5.0132838	.0000103
		.6111342

$1 - .6111342 = 0.3888658.$

From eq. (23)*, Chap. IV, we find $F(k, \phi) = 0.9226874.$ Hence

$$F(k, \phi)\left[1 - \frac{k^2}{2}\left(1 + \frac{k_0}{2} + \ldots\right)\right] = 0.3588016$$

$$\frac{k\sqrt{k_0}}{2} \sin \phi_1 = \quad 0.3290186$$

$$\frac{k\sqrt{k_0 k_{00}}}{4} \sin \phi_2 = \quad 0.0522872$$

$$\frac{k\sqrt{k_0 k_{02} k_{03}}}{8} \sin \phi_3 = -0.0013888$$

$$\frac{k\sqrt{k_0 \ldots k_{04}}}{16} \sin \phi_4 = \quad 0.0000010$$

$$\phantom{\frac{k\sqrt{k_0 \ldots k_{04}}}{16} \sin \phi_4 = \quad } 0.3799180$$

Whence

$$E(k, \phi) = 0.3588016 + 0.3799180 = 0.7387196. \quad Ans.$$

EXAMPLE. Given $k = \sin 75°$. Find $E\left(k, \dfrac{\pi}{2}\right)$.

From Example 2, Chap. XIII, we find

$$\log F\left(k, \frac{\pi}{2}\right) = 0.4421761$$

$$\log 0.3888658 = \overline{1}.5897998$$

$$\log E\left(k, \frac{\pi}{2}\right) = 0.0319759$$

$$E\left(k, \frac{\pi}{2}\right) = 1.076405. \quad Ans.$$

EXAMPLE. Given $k = \sin 30°$, $\phi = 81°$. Find $E(k, \phi)$.
    Ans. $E(k, \phi) = 1.33124.$

EXAMPLE. Find $E(\sin 80°, 55°)$.   Ans. 0.82417.

EXAMPLE. Find $E\left(\sin 27°, \dfrac{\pi}{2}\right)$.   Ans. 1.48642.

EXAMPLE. Find $E(\sin 19°, 27°)$.   Ans. 0.46946.

CHAPTER XVII.

APPLICATIONS.

RECTIFICATION OF THE LEMNISCATE.

The polar equation of the Lemniscate is $r = a\sqrt{\cos 2\theta}$, referred to the centre as the origin. From this we get

$$\frac{dr}{d\theta} = -\frac{a \sin 2\theta}{\sqrt{\cos 2\theta}};$$

whence the length of the arc measured from the vertex to any point whose co-ordinates are r and θ

$$s = \int \left\{\left(\frac{dr}{d\theta}\right)^2 + r^2\right\}^{\frac{1}{2}} d\theta = a\int \left\{\frac{\sin^2 2\theta}{\cos 2\theta} + \cos 2\theta\right\}^{\frac{1}{2}} d\theta$$

$$= a\int \frac{d\theta}{\sqrt{\cos 2\theta}} = a\int \frac{d\theta}{\sqrt{1 - 2\sin^2 \theta}}.$$

Let $\cos 2\theta = \cos^2 \phi$, whence

$$s = a\int \frac{\frac{d\theta}{d\phi}}{\cos \phi} d\phi = a\int \frac{\sin \phi \, d\phi}{\sqrt{1 - \cos^4 \phi}}$$

$$= a\int_0^\phi \frac{d\phi}{\sqrt{1 + \cos^2 \phi}} = \frac{a}{\sqrt{2}} \int_0^\phi \frac{d\phi}{\sqrt{1 - \frac{1}{2}\sin^2 \phi}}$$

$$= \frac{a}{\sqrt{2}} F\left(\frac{1}{\sqrt{2}}, \phi\right).$$

116   ELLIPTIC FUNCTIONS.

Since $r = a\sqrt{\cos 2\theta} = a\cos\phi$, the angle ϕ can be easily constructed by describing upon the axis a of the Lemniscate a semicircle, and then revolving the radius vector until it cuts this semicircle. In the right-angled triangle of which this is one side, and the axis the hypotenuse, ϕ is evidently the angle between the axis and the revolved position of the radius vector.

RECTIFICATION OF THE ELLIPSE.

Since the equation of the ellipse is $\frac{x^2}{a^2} + \frac{y^2}{b^2} = 1$, we can assume $x = a\sin\phi$, $y = b\cos\phi$, so that ϕ is the complement of the *eccentric angle*. Hence

$$s = \int \sqrt{dx^2 + dy^2} = a\int d\phi \sqrt{1 - e^2 \sin^2\phi}$$
$$= aE(e, \phi),$$

in which e, the eccentricity of the ellipse, is the modulus of the Elliptic Integral.

The length of the Elliptic Quadrant is

$$s' = aE\left(e, \frac{\pi}{2}\right).$$

EXAMPLE. The equation of an ellipse is $\frac{x^2}{16.81} + \frac{y^2}{16} = 1$; required the length of an arc whose abscissas are 1.061162 and 4.100000 : of the quadrantal arc. · *Ans.* 5.18912 ; 6.36189.

RECTIFICATION OF THE HYPERBOLA.

On the curve of the hyperbola, construct a straight line perpendicular to the axis x, and at a distance from the centre equal to the projection of b, the transverse axis, upon the asymptote, i.e. equal to $\dfrac{b^2}{\sqrt{a^2 + b^2}}$ Join the projection of the

APPLICATIONS.

given point of the hyperbola on this line with the centre. The angle which this joining line makes with the axis of x we will call ϕ. If y is the ordinate of the point on the hyperbola, then evidently

$$y = \frac{b^2 \tan \phi}{\sqrt{a^2 + b^2}},$$

and

$$x = \frac{a}{\cos \phi} \sqrt{1 - \frac{a^2 \sin^2 \phi}{a^2 + b^2}} = \frac{a}{\cos \phi} \sqrt{1 - \frac{1}{e^2} \sin^2 \phi};$$

whence

$$s = \int \sqrt{ax^2 + dy^2} = \frac{b^2}{c} \int_0^\phi \frac{d\phi}{\cos^2 \phi \sqrt{1 - \frac{1}{e^2} \sin^2 \phi}}$$

$$= \frac{b^2}{c} \int_0^\phi \frac{d\phi}{\cos^2 \phi \sqrt{1 - k^2 \sin^2 \phi}}.$$

But

$$d(\tan \phi \sqrt{1 - k^2 \sin^2 \phi}) = d\phi \sqrt{1 - k^2 \sin^2 \phi} + d\phi \frac{1 - k^2}{\sqrt{1 - e^2 \sin^2 \phi}}$$

$$- \frac{1 - k^2}{\cos^2 \phi \sqrt{1 - e^2 \sin^2 \phi}} d\phi.$$

Consequently

$$s = \frac{b^2}{c} \int_0^\phi \frac{d\phi}{\cos^2 \phi \sqrt{1 - k^2 \sin^2 \phi}}$$

$$= \frac{b^2}{c} F(k, \phi) - cE(k, \phi) + c \tan \phi \, \Delta(k, \phi)$$

$$= \frac{b^2}{ae} F\left(\frac{1}{e}, \phi\right) - aeE\left(\frac{1}{e}, \phi\right) + ae \tan \phi \, \Delta\left(\frac{1}{e}, \phi\right).$$

EXAMPLE. Find the length of the arc of the hyperbola $\dfrac{x^2}{20.25} - \dfrac{y^2}{400} = 1$ from the vertex to the point whose ordinate is $\dfrac{40}{2.05} \tan 15°$. *Ans.* 5.231184.

EXAMPLE. Find the length of the arc of the hyperbola $\dfrac{x^2}{144} - \dfrac{y^2}{81} = 100$ from the vertex to the point whose ordinate is 0.6. *Ans.* 0.6582.

www.ingramcontent.com/pod-product-compliance
Lightning Source LLC
Chambersburg PA
CBHW020122170426
43199CB00009B/594